A FIELD BOOK OF TI

A FIELD BOOK OF THE STARS

WILLIAM TYLER OLCOTT

CONTENTS

Transtextuals Team Introduction	9
INTRODUCTION.	10
SCHEME OF STUDY.	10
THE DIAGRAMS.	12
THE CONSTELLATIONS OF SPRING.	13
URSA MAJOR (er′sa mā′-jor)—THE GREAT BEAR. (Face North.)	14
URSA MINOR (er′-sa mi′-nor)—THE LITTLE BEAR. (Face North.)	16
GEMINI (jem′-i-ni)—THE TWINS. (Face West.)	18
AURIGA (â-ri′-ga)—THE CHARIOTEER. (Face Northwest.)	20
CANCER (kan′-ser)—THE CRAB. (Face West.)	22
HYDRA (hi′-dra)—THE SEA-SERPENT. (Face South and Southwest.)	23
LEO (le′o)—THE LION. (Face South.)	25
COMA BERENICES (kō′-ma ber-e-ni′-sez)—BERENICE'S HAIR.	27
CANIS MINOR (kā′-nis mi′-nor)—THE LESSER DOG. (Face West.)	29
CORVUS (kôr′-vus)—THE CROW. (Face South.)	30
CRATER (krā′-ter)—THE CUP. (Face South.)	32
METEORIC SHOWERS.	34
THE CONSTELLATIONS OF SUMMER.	35
DRACO (drā′-ko)—THE DRAGON. (Face North.)	36
LYRA (lī′-ra)—THE LYRE.	38
CYGNUS (sig′-nus)—THE SWAN, OR THE NORTHERN CROSS.	40
AQUILA (ak′-wi-lä)—THE EAGLE, AND ANTINOÜS. (Face Southeast.)	42
DELPHINUS (del-fi′-nus)—THE DOLPHIN, OR JOB'S COFFIN. (Face Southeast.)	44
SAGITTARIUS (saj-i-tā-ri-us)—THE ARCHER. (Face South.)	46
OPHIUCHUS (of-i-ū-kus)—THE SERPENT BEARER, AND SERPENS. (Face Southwest.)	48
SCORPIUS (skôr′-pi-us)—THE SCORPION. (Face South.)	50
LIBRA (lī′-bra)—THE SCALES. (Face Southwest.)	52
CORONA BOREALIS (kō-rō′nä bō-rē-a′-lis)—THE NORTHERN CROWN.	54

HERCULES (hĕr′-kū-lēz)—THE KNEELER. 56
BOÖTES (bō-ō′tēz)—THE HERDSMAN, OR BEAR DRIVER. (Face West.) 58
VIRGO (ver′-gō)—THE VIRGIN. (Face West.) 60
CANES VENATICI (kā′-nēz ve-nat′-i-cī)—THE HUNTING DOGS. (Face Northwest.) 62

METEORIC SHOWERS. **64**
JULY TO OCTOBER. 64

THE CONSTELLATIONS OF AUTUMN. **66**
CASSIOPEIA (kas-i-ō-pē′-ya)—THE LADY IN THE CHAIR. (Face North.) 66
CEPHEUS (sē′-fūs) (Face North.) 69
PEGASUS (peg′-a-sus)—THE WINGED HORSE. (Face South.) 71
ANDROMEDA (an-drom′-e-dä)—THE CHAINED LADY. 73
PERSEUS (per′-sūs)—THE CHAMPION. (Face Northeast.) 75
PISCES (pis′ēz)—THE FISHES. (Face Southeast.) 77
TRIANGULUM (trī-an′-gū-lum)—THE TRIANGLE. (Face East.) 79
AQUARIUS (a-kwā′ri-us)—THE WATER CARRIER. (Face Southwest.) 81
CAPRICORNUS (kap-ri-kôr′-nus)—THE SEA GOAT. (Face Southwest.) 83
ARIES (ā′-ri-ēz)—THE RAM. (Face Southeast.) 85
CETUS (sē′-tus)—THE WHALE. (Face Southeast.) 87
MUSCA (mus′-kä)—THE FLY. (Face Southeast.) 89

METEORIC SHOWERS. **91**
OCTOBER TO JANUARY. 91

THE CONSTELLATIONS OF WINTER. **93**
TAURUS (tâ′-rus)—THE BULL. (Face Southwest.) 93
ORION (ŏ-rī′-on)—THE GIANT HUNTER. (Face South.) 96
LEPUS (lē′-pus)—THE HARE. (Face South.) 98
COLUMBA NOACHI (co-lum′-bä nō-ä′-ki)—NOAH'S DOVE. (Face South.) 100
CANIS MAJOR (kā′-nis mā-jor)—THE GREATER DOG. (Face South.) 102
ARGO NAVIS (är′-go nā′-vis)—THE SHIP ARGO. (Face South.) 104
MONOCEROS (mō-nos′-e-ros)—THE UNICORN. (Face South.) 106
ERIDANUS (ē-rid′-a-nus)—OR THE RIVER PO. (Face Southwest.) 108

METEORIC SHOWERS. **110**

JANUARY TO APRIL.	110
THE PLANETS.	**113**
MERCURY.	114
VENUS.	114
MARS.	114
JUPITER.	115
SATURN.	115
URANUS.	116
NEPTUNE.	116
THE MILKY WAY.	**119**
THE MOTIONS OF THE STARS.	**120**
METEORS, OR SHOOTING-STARS.	**122**
NOTE.	**124**
THE NAMES OF THE STARS AND THEIR MEANINGS.	**125**

Transtextuals Team Introduction

A Field Book of the Stars takes us back to a kinder, gentler age of astronomy, where the overwhelming majority of stargazers looked at the skies with their eyes only, and those who had access to a telescope were among the lucky few.

This field book is full of charming traces of a bygone era, with its advice of suitable targets for opera glasses, and Olcott's complete innocence of the evils of rampant light pollution.

The sky *can* be enjoyed with just a pair of eyes and a sense of the grandeur of the universe and the stately progress of time and season!

- Transtextuals Team

INTRODUCTION.

Considering the ease with which a knowledge of the constellations can be acquired, it seems a remarkable fact that so few are conversant with these time-honored configurations of the heavens. Aside from a knowledge of "the Dipper" and "the Pleiades," the constellations to the vast majority, are utterly unknown.

To facilitate and popularize if possible this fascinating recreation of star-gazing the author has designed this field-book. It is limited in scope solely to that purpose, and all matter of a technical or theoretical nature has been omitted.

The endeavor has been to include in these pages only such matter as the reader can observe with the naked eye, or an opera-glass. Simplicity and brevity have been aimed at, the main idea being that whatever is bulky or verbose is a hindrance rather than a help when actually engaged in the observation of the heavens.

The constellations embraced in this manual are only those visible from the average latitude of the New England and Middle States, and owe their place in the particular season in which they are found to the fact that in that season they are favorably situated for observation.

With this brief explanatory note of the purpose and design of the book, the author proceeds to outline the scheme of study.

SCHEME OF STUDY.

The table of contents shows the scheme of study to be pursued, and to facilitate the work it is desirable that the student follow the therein circumscribed order.

A knowledge on the part of the reader of Ursa Major, or "the Dipper" as it is commonly called, and "the Pleiades," the well-known group in Taurus, is presupposed by the author.

With this knowledge as a basis, the student is enabled in any season to take up the study of the constellations. By following out the order dictated, he will in a few nights of observation be enabled to identify the various configurations making up the several constellations that are set apart for study in that particular season.

A large plate, showing the appearance of the heavens at a designated time on the first night of the quarter, is inserted before each season's work. This should be consulted by the student before he makes an observation, in order that he may obtain a comprehensive idea of the relative position of the constellations, and also know in what part of the heavens to locate the constellation which he wishes to identify.

A knowledge of one constellation enables the student to determine the position of the next in order. In this work, the identification of each constellation depends on a knowledge of what precedes, always bearing in mind the fact that each season starts as a new and distinct part to be taken by itself, and has no bearing on that which comes before.

THE DIAGRAMS.

The diagrams, it will be observed, are grouped under the seasons, and they indicate the positions of the constellations as they appear at 9 o'clock P.M. in mid-season.

To facilitate finding and observing the constellations, the student should face in the direction indicated in the text. This applies to all constellations excepting those near the zenith.

The four large plates are so arranged that the observer is supposed to be looking at the southern skies. By turning the plate about from left to right, the eastern, northern, and western skies are shown successively.

On many of the diagrams the position of nebulæ is indicated. These are designated by the initial letter of the astronomer who catalogued them, preceded by his catalogue number, as for instance 8 M. signifies nebula number 8 in Messier's catalogue.

The magnitudes assigned to the stars in the diagrams are derived from the Harvard Photometry. When a star is midway between two magnitudes the numeral is underlined, thus *2*, indicates a star of magnitude 2.5.

If a star's magnitude is between 1 and 1.5 it is regarded as a first-magnitude star. If it lies between 1.5 and 2 it is designated second magnitude.

THE CONSTELLATIONS OF SPRING.

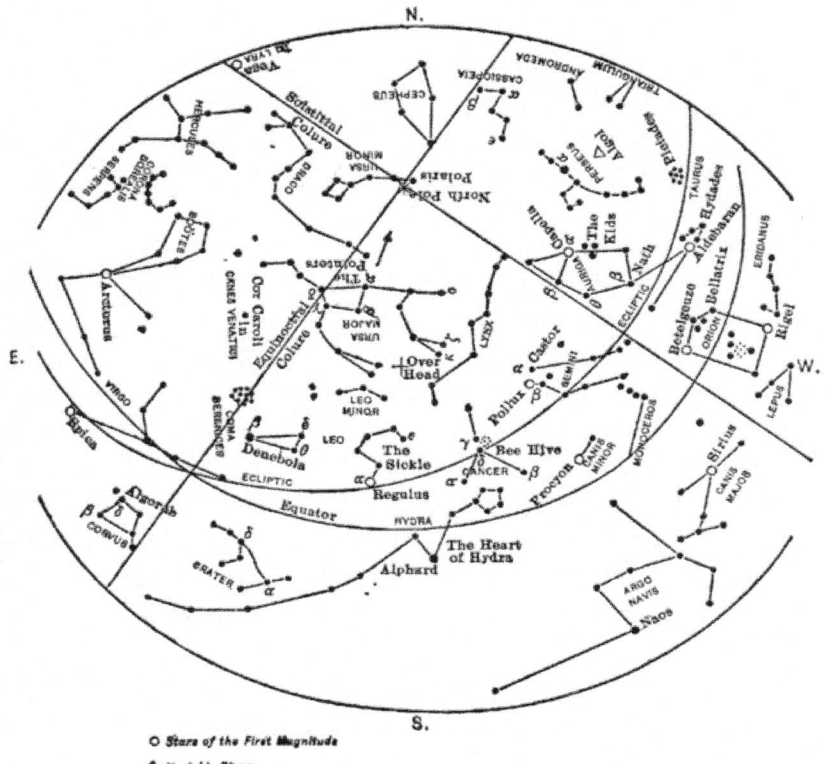

URSA MAJOR (er′sa mā′-jor)—THE GREAT BEAR. (Face North.)

LOCATION.—Ursa Major is probably the best known of the constellations, and in this work I presuppose that the reader is familiar with its position in the heavens. It is one of the most noted and conspicuous constellations in the northern hemisphere, and is readily and unmistakably distinguished from all others by means of a remarkable cluster of seven bright stars in the northern heavens, forming what is familiarly termed "The Dipper."

The stars α and β are called the pointers, because they always point toward the Pole Star, $28\frac{3}{4}°$ distant from α.

Alioth is very nearly opposite Shedir in Cassiopeia, and at an equal distance from the Pole. The same can be said of Megres, in Ursa Major, and Caph, in Cassiopeia.

The star o is at the tip of the Bear's nose. A clearly defined semicircle begins at o and ends in the pair ι and κ at the extremity of the Bear's right fore paw. This group of stars resembles a sickle. Note little Alcor close to Mizar. This star was used by the Arabs as a test of good eyesight.

Mizar and Alcor are known as the horse and his rider.

This plate shows the Bear lying on his back, his feet projected up the sky; three conspicuous pairs of stars represent three of his four feet.

The Chaldean shepherds and the Iroquois Indians gave to this constellation the same name. The Egyptians called it "The Thigh." α and η are moving through space in a contrary direction to the remaining five stars in "The Dipper."

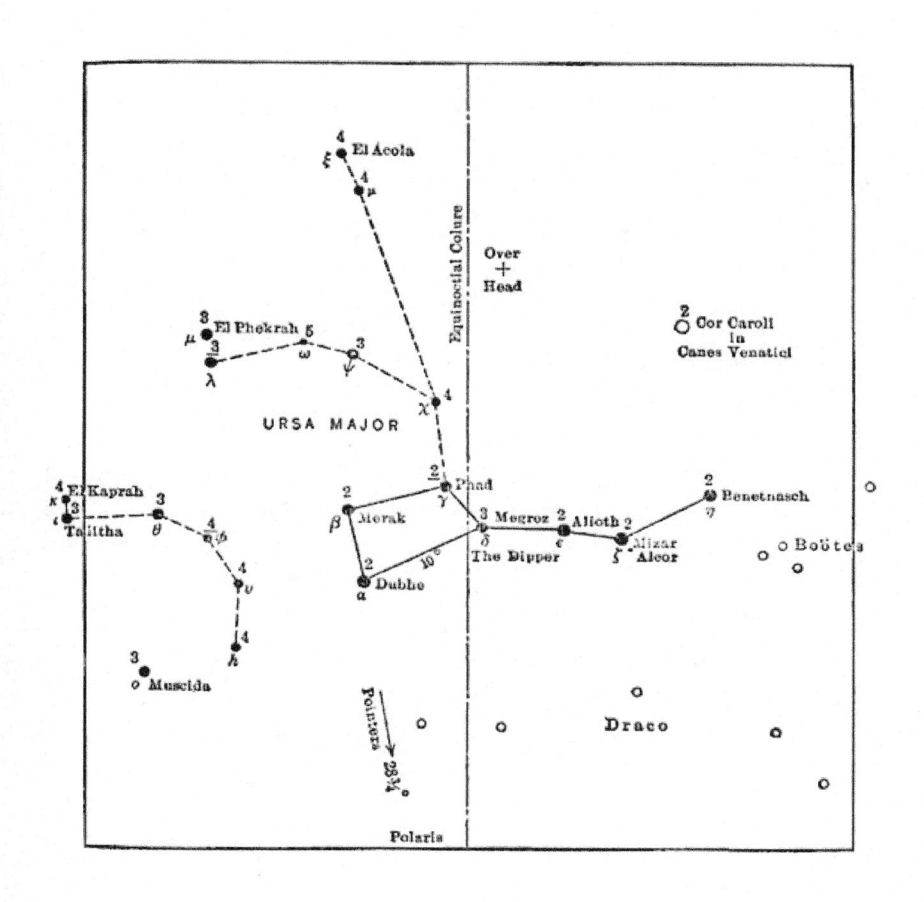

URSA MAJOR

URSA MINOR (er'-sa mi'-nor)—THE LITTLE BEAR. (Face North.)

LOCATION.—The two pointer stars in Ursa Major indicate the position of Polaris, the North Star, which represents the tip of the tail of the Little Bear, and the end of the handle of the "Little Dipper." In all ages of the world, Ursa Minor has been more universally observed and more carefully noticed than any other constellation, on account of the importance of the North Star.

Polaris is a little more than 1¼° from the true pole. Its light takes fifty years to reach us.

A line joining β Cassiopeiæ, and Megres, in Ursa Major, will pass through Polaris.

At the distance of the nearest fixed star our sun would shine as a star no brighter than Polaris which is presumably about the sun's size.

Polaris revolves around the true pole once in twenty-four hours in a little circle 2½° in diameter. Within this circle two hundred stars have been photographed.

The North Star is always elevated as many degrees above the horizon as the observer is north of the equator.

Compare the light of the four stars forming the bowl of the "Little Dipper," as they are each of a different magnitude. A standard first-magnitude star is 2½ times brighter than a standard second magnitude star, etc.

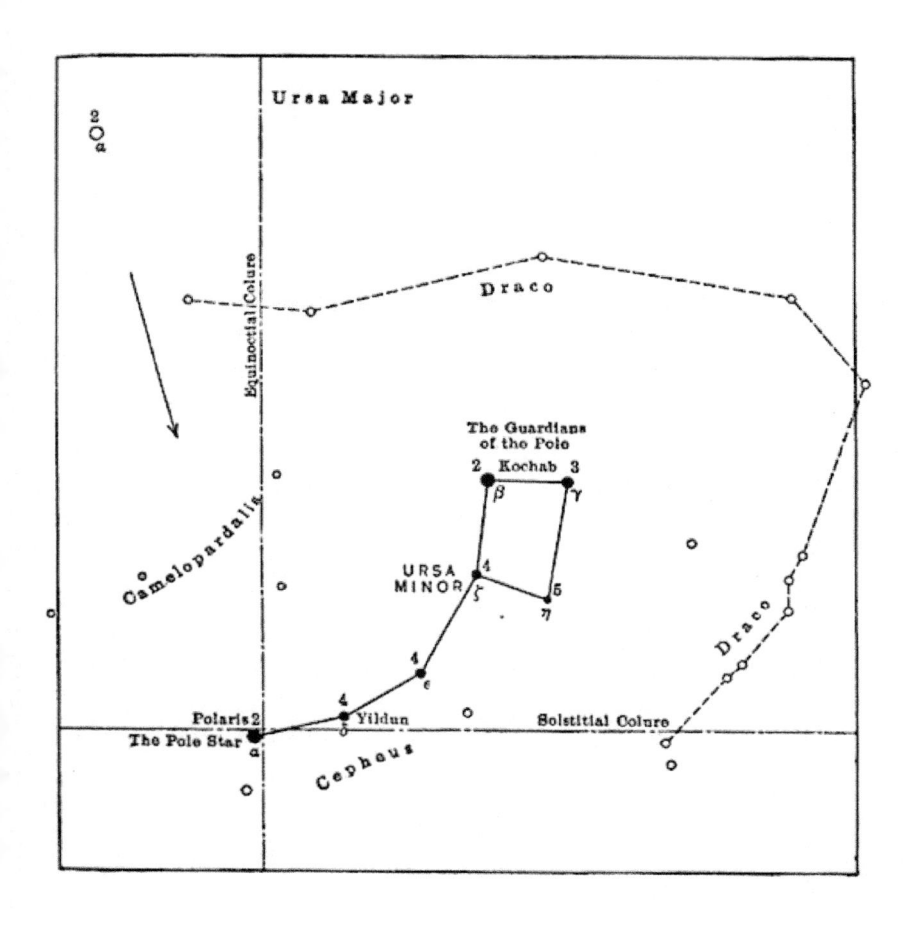

URSA MINOR

GEMINI (jem´-i-ni)—THE TWINS. (Face West.)

LOCATION.—A line drawn from β to κ Ursæ Majoris and prolonged an equal distance ends near Castor, in Gemini. Gemini is characterized by two nearly parallel rows of stars. The northern row if extended would reach Taurus, the southern one Orion. Note the fine cluster 35 M. Herschel discovered Uranus in 1781 a short distance southwest of it. Two wonderful streams of little stars run parallel northwest on each side of the cluster. Where the ecliptic crosses the solstitial colure is the spot where the sun appears to be when it is farthest north of the equator, June 21st. Castor is a fine double for a telescope, and Pollux has three little attendant stars. An isoceles triangle is formed by Castor, Aldebaran in Taurus, and Capella in Auriga. There is a record of an occultation in Gemini noted about the middle of the fourth century B.C.

The Arabs saw in this group of stars two peacocks, the Egyptians two sprouting plants, and the Hindus twin deities, while in the Buddhist zodiac they represented a woman holding a golden cord. Since classic times, however, the figure has always been that of human twins.

At the point indicated near θ a new star was discovered by Enebo in March, 1912. It attained a maximum of about magnitude 3.5 and has at this writing waned to the eleventh magnitude.

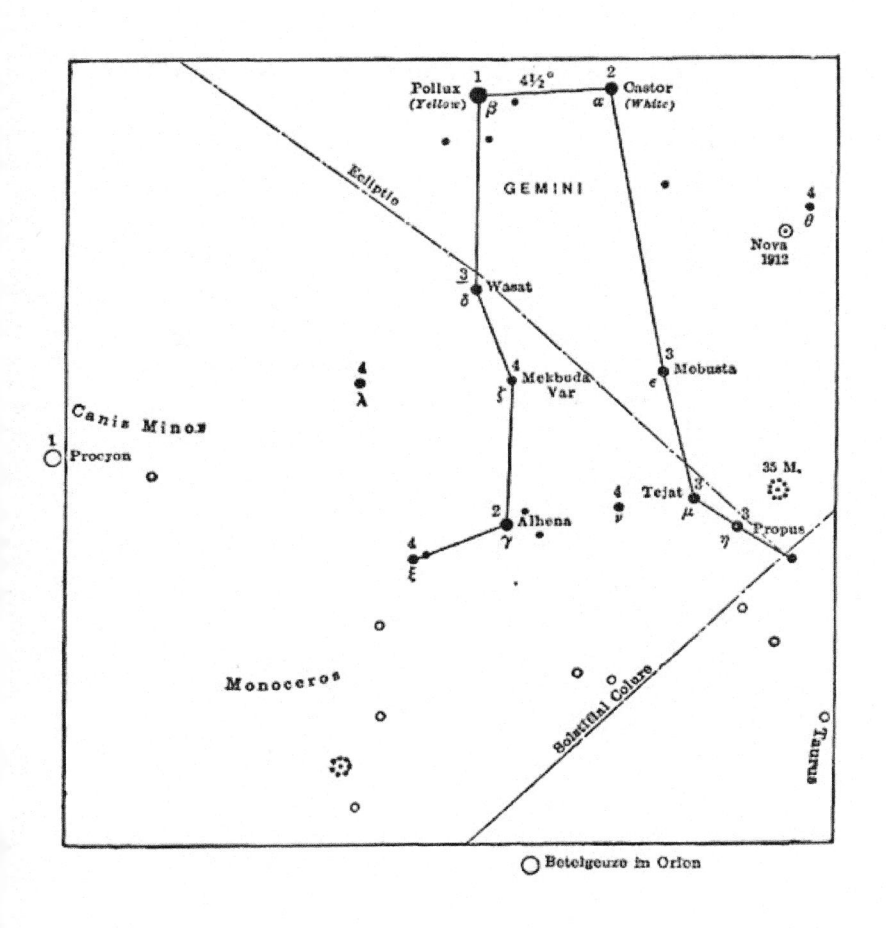

GEMINI

AURIGA (â-rī′-ga)—THE CHARIOTEER. (Face Northwest.)

LOCATION.—A line drawn from δ to α Ursæ Majoris, and prolonged about 45°, ends near the bright Capella, in Auriga, a star of the first magnitude, and one of the most brilliant in the heavens. It is unmistakable, having no rival in brightness near it. Auriga is a beautiful and conspicuous constellation. It is characterized by a clearly defined pentagon. Note the three fourth-magnitude stars near Capella known as "The Kids." The star β is common to Auriga and Taurus, being the former's right foot and the latter's northern horn. The field within the pentagon is particularly rich in clusters. Capella forms a rude square with Polaris, ε Cassiopeiæ, and o Ursæ Majoris, and forms an equilateral triangle with Betelgeuze in Orion, and the Pleiades in Taurus.

A line from θ to α Aurigæ prolonged about 20° ends near α Persei.

Capella is visible at some hour of every clear night throughout the year. Of the first-magnitude stars it is nearest to the Pole, and it rises almost exactly in the northeast.

To the Arabs Capella was "The Driver," because it seemed to rise earlier than the other stars and so apparently watched over them, or still more practically as "The Singer" who rode before the procession cheering on the camels, which last were represented by the Pleiades.

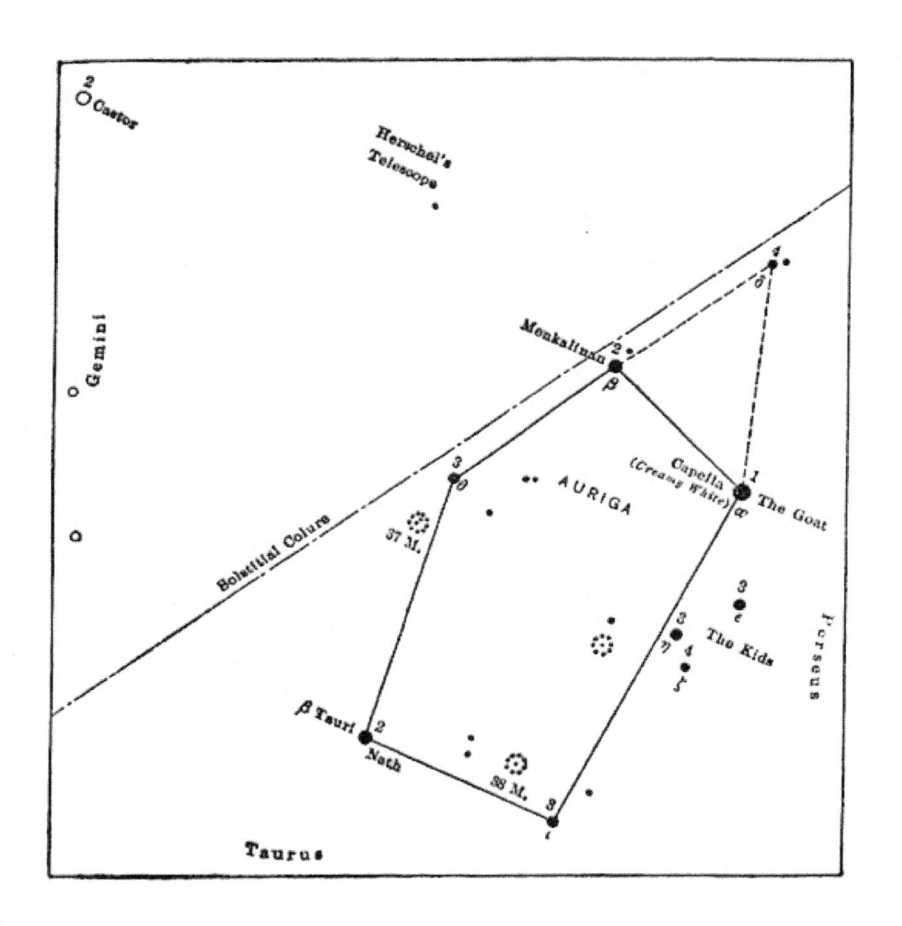

AURIGA

CANCER (kan´-ser)—THE CRAB. (Face West.)

LOCATION.—Cancer lies between Gemini and Leo. A line drawn from Nath in Auriga to Pollux in Gemini, and prolonged about 15°, ends in Præsepe, the Manger, the great star cluster in Cancer, which is also called "The Bee Hive." It contains 300 stars. The stars γ and δ are called the Aselli—the ass's colts feeding from the silver manger.

The star β lies about 10° northeast of Procyon. Acubens, α lies on the same line the same distance beyond β. These two stars form the tips of the inverted "Y" which distinguishes Cancer.

An imaginary line from Capella through Pollux will point out Acubens. Close to it are two faint stars. The Bee Hive lies within an irregular square formed by γ, δ, η, and θ, and looks like a nebula to the naked eye.

In June, 1895, all the planets except Neptune were in this quarter of the heavens, and Halley's comet was in this constellation on its first appearance in 1531.

The dimness of γ and δ is an infallible precursor of rain, and if the Bee Hive is not visible in a clear sky, it is a presage of a violent storm.

HYDRA (hi´-dra)—THE SEA-SERPENT. (Face South and Southwest.)

LOCATION.—The head of Hydra, a striking and beautiful arrangement of stars, lies just below the Bee Hive, in Cancer, 6° south of Acubens in that constellation, and forms a rhomboidal figure of five stars.

Hydra is about 100° in length and reaches almost from Canis Minor to Libra. Its stars are all faint except Alphard, or the Hydra's heart, a second-magnitude star remarkable for its lonely situation, southwest of Regulus, in Leo. A line drawn from γ Leonis through Regulus points it out. It is of a rich orange tint. Castor and Pollux, in Gemini, point southeast to it.

The constellations Crater, the Cup, and Corvus, the Crow, both stand on the coils of Hydra, south of Denebola, the bright star in the tail of the Lion.

Hydra is supposed to be the snake shown on a uranographic stone from the Euphrates, 1200 B.C.

The little asterism Sextans, the Sextant, lies in the region between Regulus and Alphard. It contains no stars brighter than the fourth magnitude.

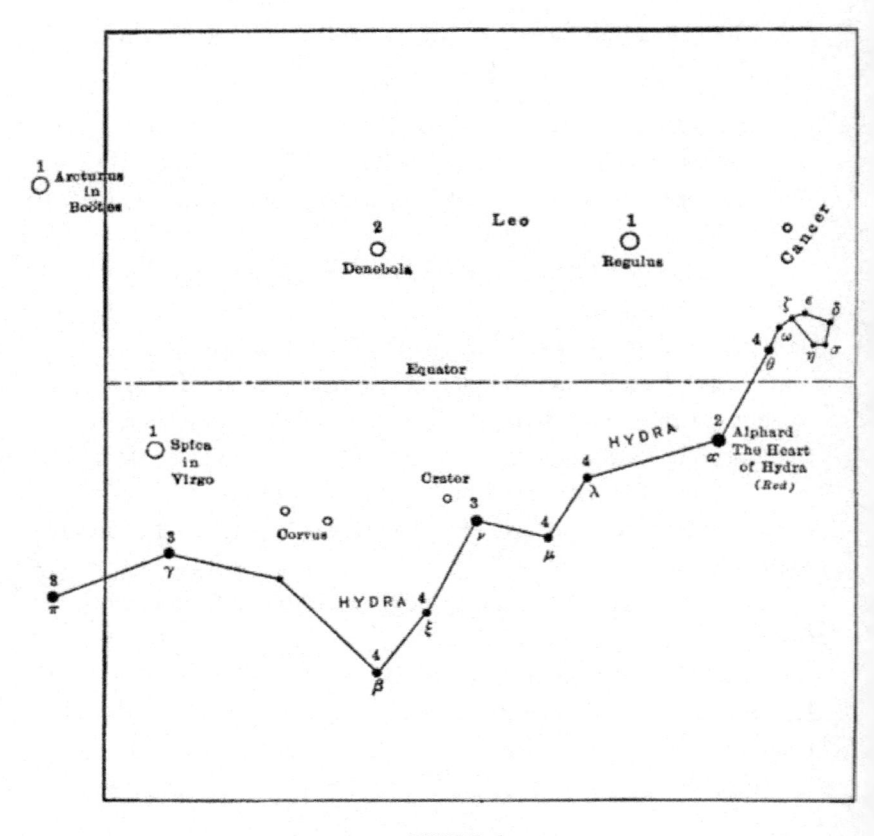

HYDRA

LEO (le'o)—THE LION. (Face South.)

LOCATION.—A line drawn from Pollux, in Gemini, to γ in Cancer, and prolonged about 12°, strikes Regulus, the brilliant star in the heart of the Lion. Regulus lies about 9° east of Acubens, in Cancer, and about 12° northeast of Alphard, in the heart of Hydra.

Leo is one of the most beautiful constellations in the zodiac. It lies south of the Great Bear, and its principal stars are arranged in the form of a sickle which nearly outlines the Lion's head. This group is so striking as to be unmistakable. Regulus is in the handle of the sickle. It is one of the stars from which longitude is reckoned, lies almost exactly on the ecliptic, and is visible for eight months in the year.

Denebola, the bright star in the Lion's tail, lies 25° east of Regulus, and about 35° west of Arcturus, in Boötes. It is the same distance northwest of Spica, in Virgo, and forms with Spica and Arcturus a large equilateral triangle.

ζ is double, and has three faint companion stars.

ε has two seventh-magnitude companion stars, forming a beautiful little triangle.

Regulus is white in color, γ yellow, π red.

γ is a beautiful colored telescopic double star and has a companion visible in an opera-glass.

The figure of Leo very much as we now have it appears in all the Indian and Egyptian zodiacs.

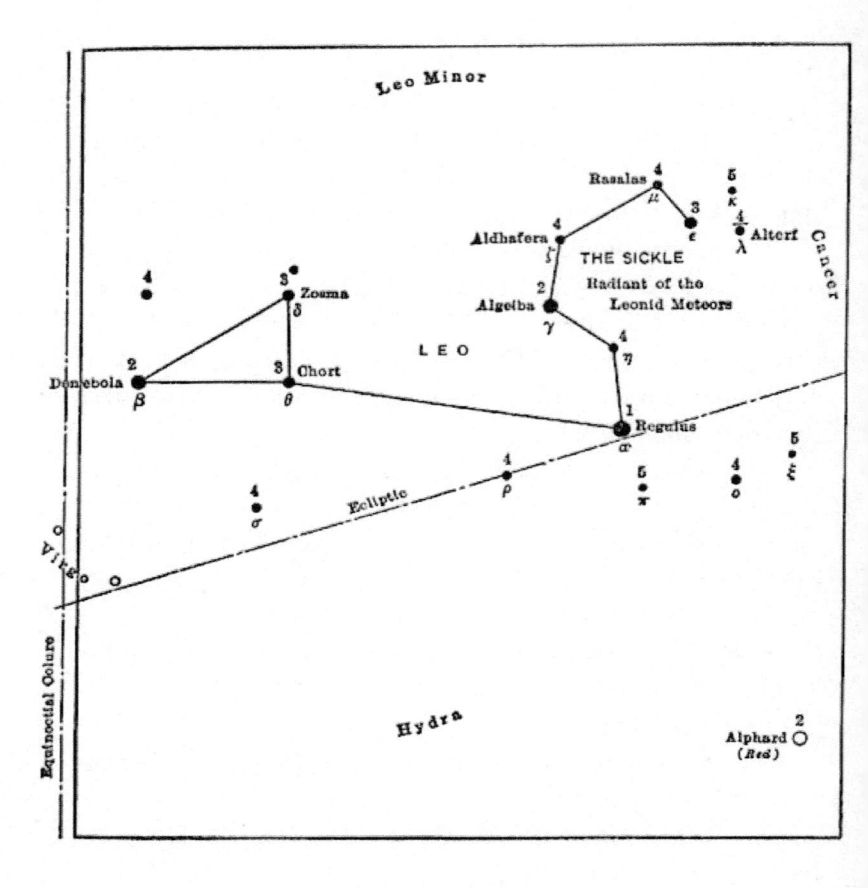

LEO THE SICKLE

COMA BERENICES (kō′-ma ber-e-ni′-sez)—BERENICE'S HAIR.

LOCATION.—A line drawn from Regulus to Zosma, in Leo, and prolonged an equal distance, strikes this fine cluster, which is 18° northeast of Zosma, δ Leonis.

The group lies well within a triangle formed by Denebola, Arcturus, in Boötes, and Cor Caroli, in Canes Venatici, which triangle is the upper half of the Diamond of Virgo.

Twenty or thirty stars in this group can be counted with an opera-glass, and the group can be easily distinguished with the naked eye, when the moon is not visible.

The first half of the month of April can be called the most brilliant sidereal period of the year. At this time eleven first-magnitude stars are visible in this latitude at 9 P.M. From east to west they are: Vega, Arcturus, Spica, Regulus, Pollux, Procyon, Sirius, Capella, Aldebaran, Betelgeuze, and Rigel, truly a glorious company, an incomparable sight.

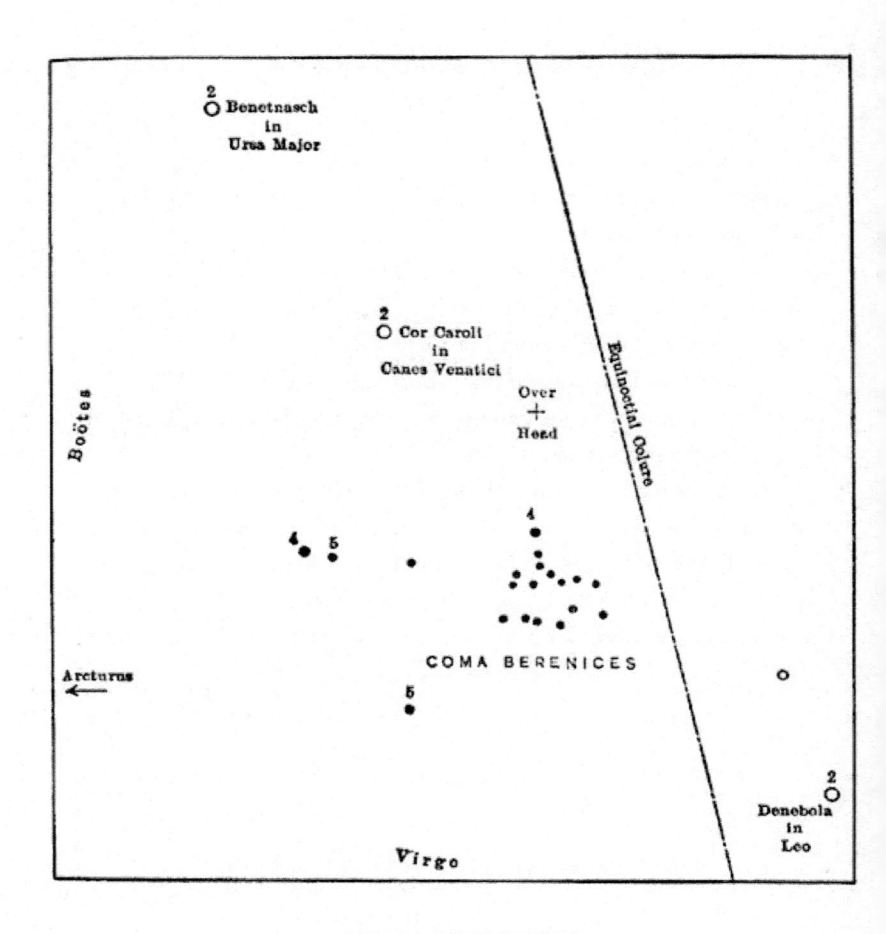

COMA BERENICES

CANIS MINOR (kā´-nis mĭ´-nor)—THE LESSER DOG. (Face West.)

LOCATION.—Procyon, the Little Dog Star, lies about 23° south of Pollux, in Gemini. A line drawn from Nath, in Auriga, to Alhena, in Gemini, and prolonged about 18°, reaches Procyon.

Procyon is equidistant from Betelgeuze in Orion, and Sirius in Canis Major, and forms with them an equilateral triangle. It forms a large right-angled triangle with Pollux and Betelgeuze.

The light from Procyon is golden yellow. Four degrees northwest of it is the third-magnitude star Gomeisa. The glass shows two small stars forming a right-angled triangle with it.

Procyon was distinctly mentioned by Ptolemy. It rises in this latitude a little north of east about half an hour before Sirius, the Dog Star, hence it was called Procyon from two Greek words which signify "before the dog."

Procyon is one of our nearest neighbors in space, at a distance of ten light years, and is attended by a very faint companion which is only visible in the largest telescopes.

CORVUS (kôr´-vus)—THE CROW. (Face South.)

LOCATION.—A line drawn from the Bee Hive, in Cancer, through Regulus, in Leo, and prolonged about 40°, ends near the conspicuous quadrilateral which distinguishes Corvus. The brightest star in this region of the sky is Spica, in Virgo. It lies about 10° northeast of Algorab.

ζ is a double star for an opera-glass. A faint pair of stars lie close below and to the west of β. The Crow is represented as standing on, and pecking at, the coils of Hydra. The star Al Chiba is in the Crow's bill.

Corvus was known as the Raven in Chaucer's time.

δ is an interesting telescopic double.

A line drawn from γ to β Corvi and prolonged twice its length locates the third-magnitude star ι Centauri in the right shoulder of the Centaur. The brightest stars in this constellation are not visible in this latitude.

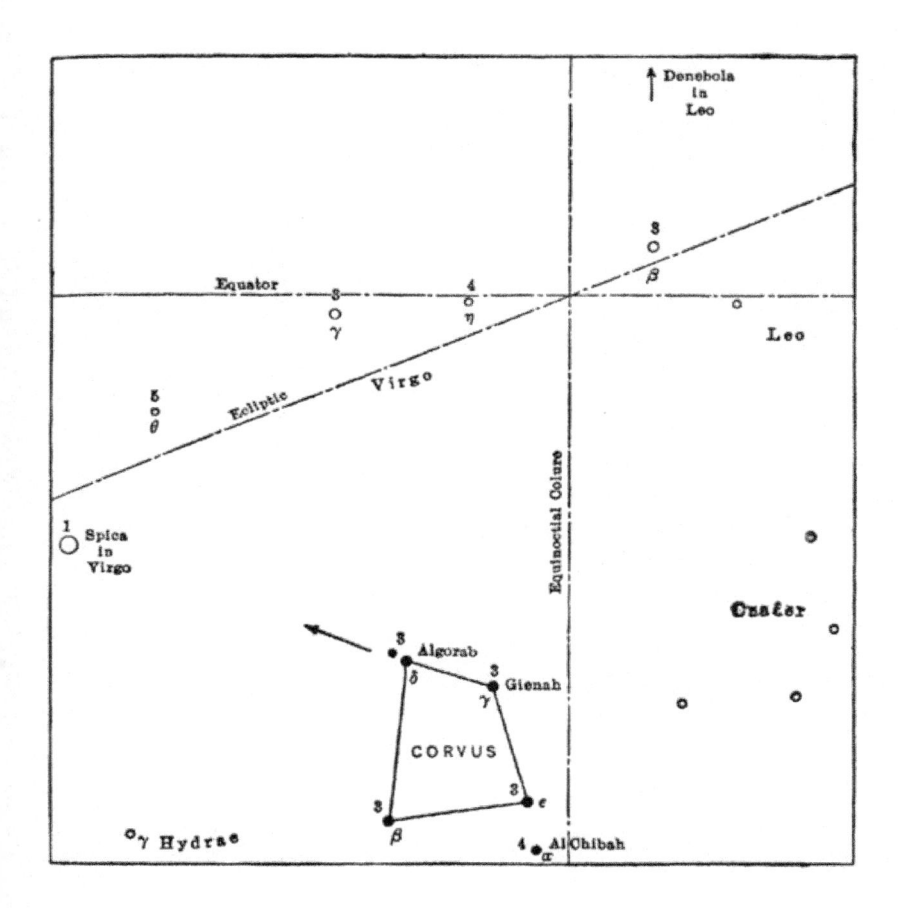

CORVUS

CRATER (krā'-ter)—THE CUP. (Face South.)

LOCATION.—Crater is situated 15° west of Corvus, and due south of θ Leonis. It is easily distinguished by reason of a beautiful and very striking semicircle of six stars of the fourth magnitude, forming the bowl of the cup.

The constellation resembles a goblet with its base resting on the coils of Hydra.

The star Alkes is common to Hydra and Crater, and may be seen 24° southeast of Alphard in the heart of Hydra. It is distinguished by its forming an equilateral triangle with α and γ, stars of the same magnitude 6° south and east of it.

Corvus and Crater are to be seen half-way up the southern sky during the early evenings in spring.

δ is now the lucida.

Crater is situated at about the centre of Hydra and is on the meridian, April 26th. Owing to its many faint stars it is best seen on a clear moonless night.

The zodiacal light is well worth observing at this season of the year. It is to be seen in the western sky shortly after sundown, and is most intense during the evenings of March.

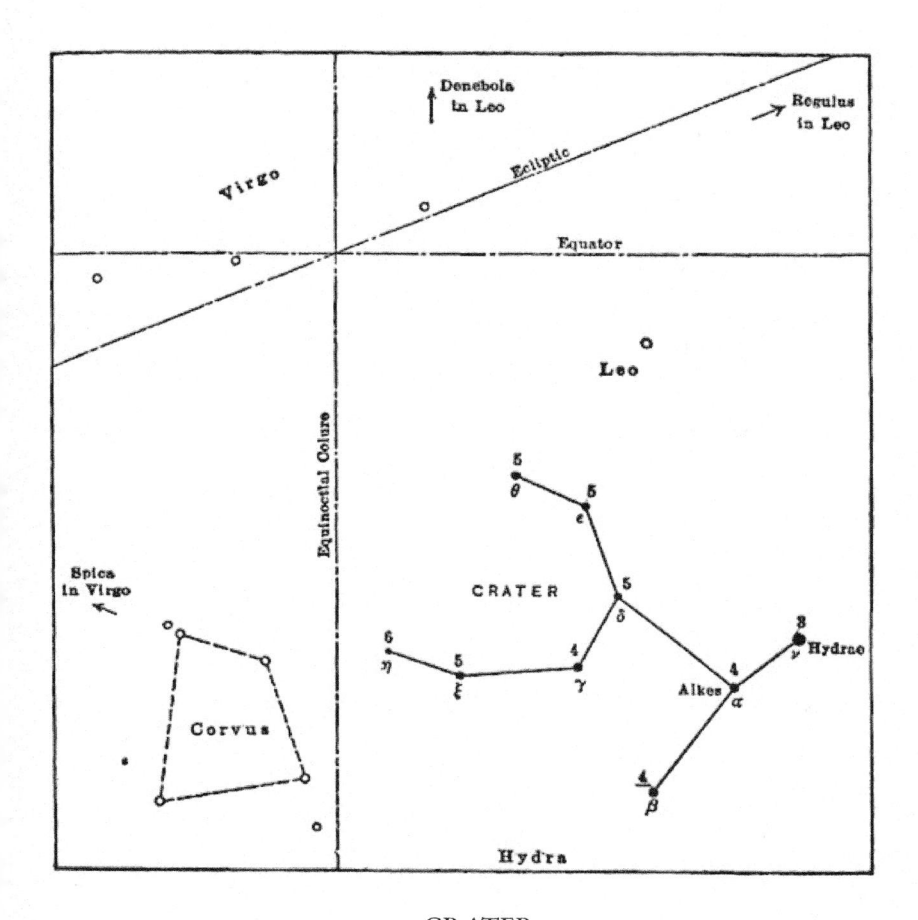

CRATER

METEORIC SHOWERS.

APRIL TO JULY.

Name of Shower	Date	Radiant Point	Characteristics	Other Dates of Observation	Location
Beta or Mu Draconids Beta Serpentids	Apr. 9-16 Apr. 18	The Dragon's head The Serpent's head	Sw. F.	Apr. 17-25	N.E. S.E.
Lyrids, rich shower	Apr. 20	About 10° from Vega toward Hercules	V. Sw.		N.E.
Eta Aquarids, fine annual shower	May 6	Near the Water Jar	Sw. Sk.	After 2 A.M.	E.
Alpha Coronids, well defined in 1885	May 11	Near Gemma (α) Coronæ B.	Sl. F.	May 7-18	N.
Iota Pegasids, well defined shower	May 30	Between Cygnus and the Great Square	Sw. Sk.	May 29-June 4 after 10 P.M.	N.E.
Beta Herculids Beta Ophiuchids	June 7 June 10	Near the Crown About 8°S. of Ras Alhague S.E.	Sl. B. Sl.	A fire ball radiant June 10, 13	S.E.
Delta Cepheids	June 20	About 13° from (β) Cassiopeiæ	Sw.	June 10-28, July 19, Aug. 25, etc.	N.

The Abbreviations under *Characteristics* are as follows:

V.—very.
M.—moderately.
Sw.—swift.
Sl.—slow.
Sh.—short.

B.—bright.
F.—faint.
Sk.—streak-leaving meteors.
T.—train-leaving meteors

THE CONSTELLATIONS OF SUMMER.

DRACO (drā'-ko)—THE DRAGON. (Face North.)

LOCATION.—About 10° from α Ursæ Majoris—from α to δ is 10°—slightly south of, that is above, the line from α to Polaris, is Giansar, λ in the tip of the Dragon's tail. Above λ, and almost in line with it, are two more stars in Draco, which form with two stars in Ursa Major a quadrilateral. (See diagram.) Draco now curves sharply eastward, coiling about the Little Bear as shown, then turns abruptly southerly, ending in a characteristic and clearly defined group of four stars, forming an irregular square, representing the Dragon's head. This group is almost overhead in the early evening in summer. The star in the heel of Hercules lies just south of the Dragon's head. The brilliant Vega will be seen about overhead, 12° southwest of the Dragon's head. Eltanin, one of the Dragon's eyes, is noted for its connection with the discovery of the law of aberration of light. It is of an orange hue, while the star β, near it, is white. Note Thuban, once the Pole Star, at one corner of a quadrilateral that Draco forms with Ursa Major.

Thuban could be seen by day or night from the bottom of the central passage of several of the Pyramids in Egypt.

The rising of Eltanin was visible about thirty-five hundred years B.C. through the central passages of the temples of Hathor at Denderah. The Egyptians called Draco "The Hippopotamus."

Vega and the four stars in the Dragon's head offer an opportunity to compare the first five stellar magnitudes with which all should be familiar.

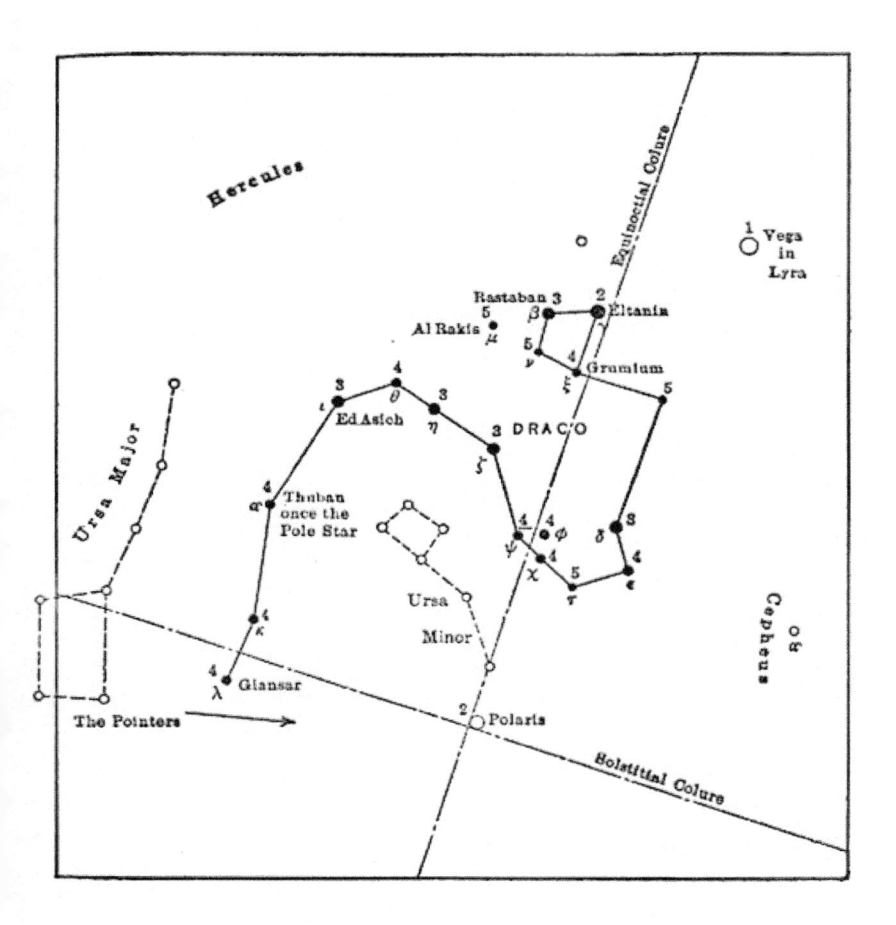

DRACO

LYRA (lī'-ra)—THE LYRE.

LOCATION.—Lyra may be easily distinguished because of the brilliant Vega, its brightest star, which is situated about 12° southwest of the Dragon's head. It is unmistakable, as it is the brightest star in this region of the heavens, and the third brightest in this latitude. In July and August Vega is close to the zenith in the early evening.

The six bright stars in Lyra form an equilateral triangle on one corner of a rhomboid. A very characteristic figure.

ε is a pretty double for an opera-glass, and a 3" glass reveals the duplicity of each star of this pair. ε is therefore a double double.

ζ is a double for a good glass.

β is a variable, changing from magnitude 3.4 to 4.4 in twelve days. At its brightest it is about equal to its near neighbor γ Lyræ.

The noted ring nebula lies between β and γ. A 3" glass reveals it but a powerful telescope is required to render its details visible.

If the distance from the earth to the sun equalled one inch, the distance from the earth to Vega would be 158 miles.

Vega was the first star to be photographed, in 1850. It is visible at some hour every clear night, and has been called the arc-light of the sky. Its light has the bluish-white hue that suggests "a diamond in the sky."

The spectroscope reveals that Vega is a star probably only in its infancy, as hydrogen is its predominating element.

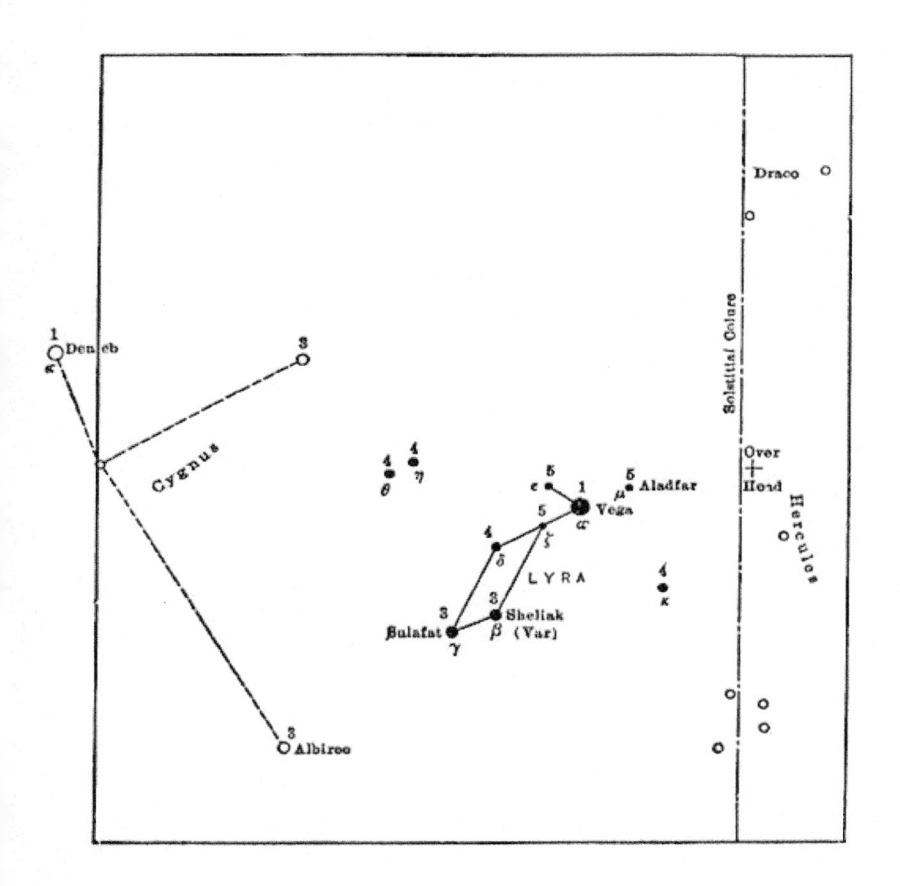

LYRA

CYGNUS (sig´-nus)—THE SWAN, OR THE NORTHERN CROSS.

LOCATION.—Deneb, the brightest star in Cygnus, is at the top of the cross, and a little over 20° east of Vega. It forms a triangle with Vega and Altair in Aquila—Altair being at the apex, about 35° from Deneb and Vega.

β Cygni is at the base of the cross, and a line drawn from Vega to Altair nearly touches it. It is a beautiful colored double for a small telescope.

Note "61," one of the nearest stars to us. It was the first star whose distance was measured (by Bessel in 1838). It is a double star and 10.4 light years distant.

The cross is nearly perfect and easily traced out. It lies almost wholly in the Milky Way.

Note "The Coal Sack," one of the dark gap in the Milky Way.

Cygnus contains an unusual number of deeply colored stars and variable stars.

o Cygni has a sixth-magnitude companion, and γ is in the midst of a beautiful stream of faint stars.

This region is perhaps richer than any similar extent in the heavens. An opera-glass will reveal many of its beauties.

Herschel counted 331,000 stars in an area of only 5° in Cygnus.

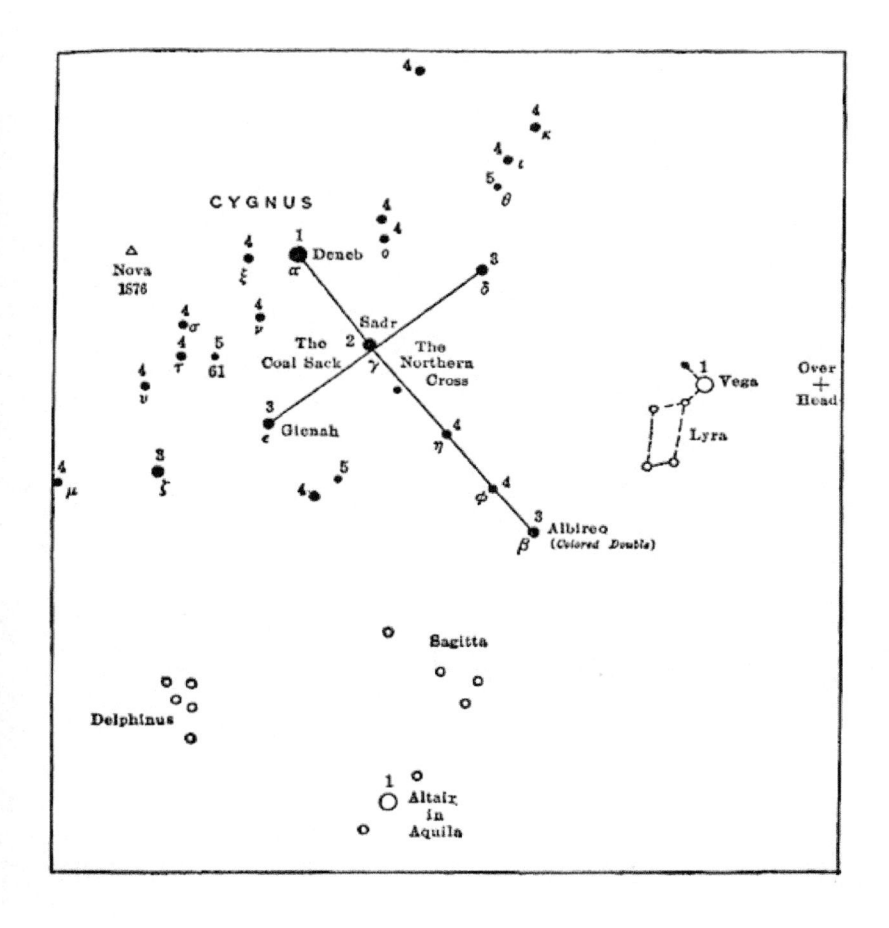

CYGNUS

AQUILA (ak´-wi-lä)—THE EAGLE, AND ANTINOÜS.
(Face Southeast.)

LOCATION.—Half-way up the sky in the Milky Way, you will see three stars in a line, the middle one much brighter than the other two. This bright star is Altair, in Aquila. It forms with Vega and Deneb an isosceles triangle. Altair is at the apex, about 35° from the other two. A triangle is formed by Vega, Altair, and Ras Alhague, in the Serpent Bearer, which is about 30° west of Altair.

This is a double constellation composed of Aquila and Antinoüs. Altair is in the neck of the Eagle, Alschain in the head of Antinoüs.

When the moon is absent, a rude arrowhead can be traced out, embracing almost all the stars in Aquila.

η is an interesting variable star, changing from magnitude 3.5 to 4.7 and back again within a period of 7 days 4 hours 12 minutes.

Altair rises about 8° north of the exact eastern point on the horizon.

In A.D. 389 a wonderful temporary star flashed out near Altair that equalled Venus in brightness and vanished within three weeks' time.

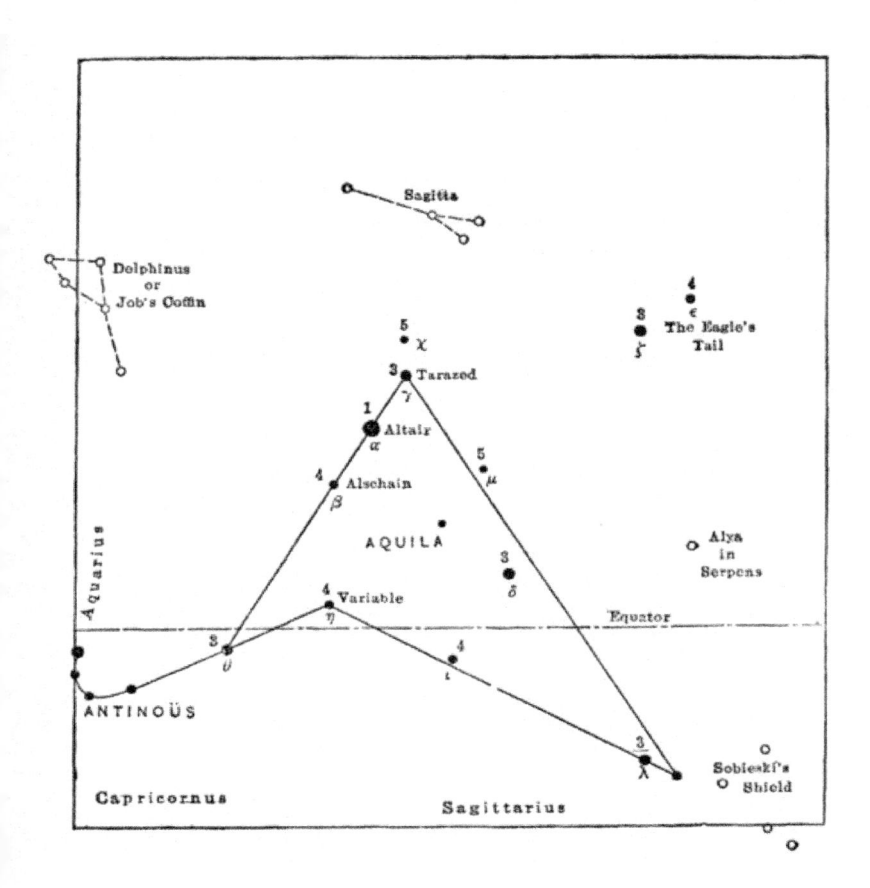

AQUILA & ANTINOÜS

DELPHINUS (del-fī'-nus)—THE DOLPHIN, OR JOB'S COFFIN. (Face Southeast.)

LOCATION.—The little cluster of five stars forming Delphinus is to be seen about 10° northeast of Altair, and, though there are no bright stars in the group, it can hardly escape notice. A line drawn from Vega to Albireo, and prolonged about 20°, strikes the star ε in the tail of the Dolphin. The four other stars of prominence in the constellation are a little above ε, and form a diamond-shaped figure.

The little asterisms Sagitta, the Arrow, and Vulpecula and Anser, the Fox and Goose, are shown just above Delphinus.

Delphinus is also called Job's Coffin. The origin of this appellation is unknown.

In Greece, Delphinus was the Sacred Fish, the sky emblem of philanthropy. The Arabs called it the "Riding Camel."

The star γ Delphini is a fine double for a small telescope with a marked and beautiful contrast of colors.

The names for α and β reversed spell "Nicolaus Venator," the Latinized name of the assistant to the astronomer Piazzi.

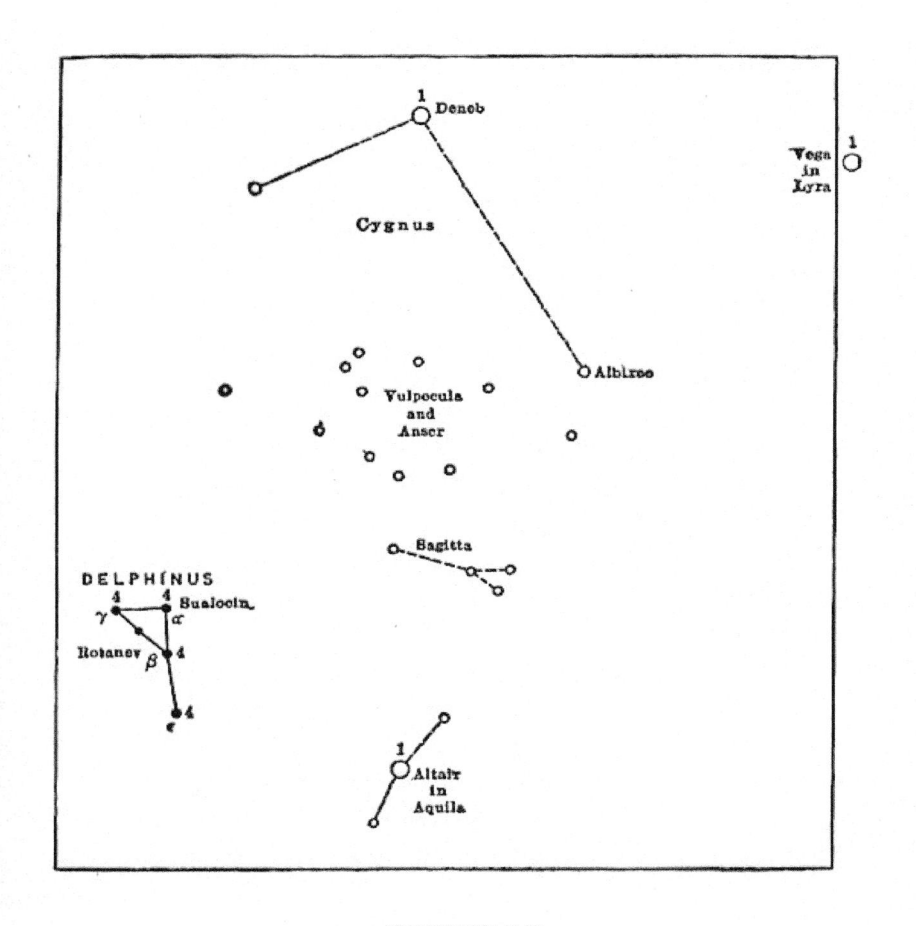

DELPHINUS

SAGITTARIUS (saj-i-tā-ri-us)—THE ARCHER. (Face South.)

LOCATION.—A line drawn from Deneb, in Cygnus, to Altair, in Aquila, and prolonged an equal distance, terminates in Sagittarius about 10° east of its distinguishing characteristic, the Milk Dipper. Sagittarius is one of the signs of the zodiac, and lies between Capricornus, on the east, and Scorpius, on the west.

The bow is easily traced out. γ marks the arrow's tip.

Note the star μ, which serves to point out the Winter Solstice, where the solstitial colure intersects the ecliptic.

On a clear night, the pretty cluster known as Corona Australis, the Southern Crown, can be seen about 10° below the bowl of the Milk Dipper. Its lucida, the fourth-magnitude star Alfecca Meridiana culminates at 9 P.M., August 13th.

Sagittarius is about due south, in a splendid position for observation, during the month of July, between the hours of nine-thirty and eleven o'clock P.M.

Observe with an opera-glass the fine clusters 20 M. and 8 M., also an almost circular black void near the stars γ and δ, and to the east of this spot another of narrow crescent form.

The stars φ and ζ in the Milk Dipper are moving in opposite directions. Future generations therefore will not have this time-honored figure to guide them in locating the Archer in their summer night skies.

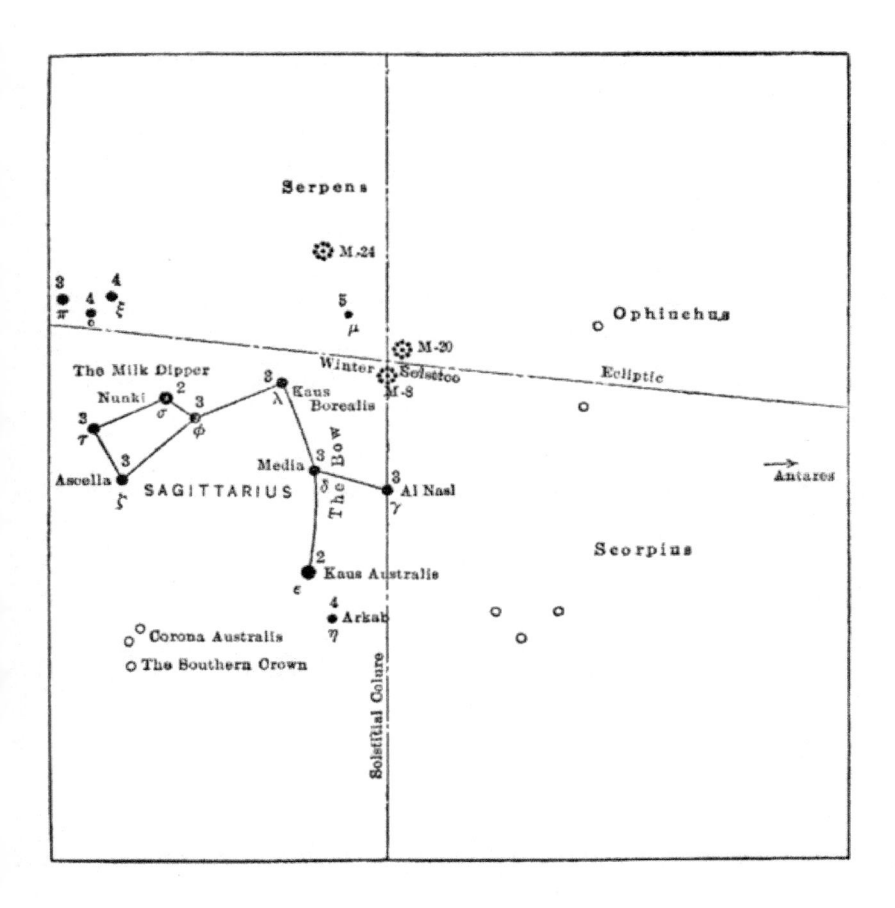

SAGITTARIUS

OPHIUCHUS (of-i-ū-kus)—THE SERPENT BEARER, AND SERPENS. (Face Southwest.)

LOCATION.—A line drawn from ε Delphini to γ Aquilæ, prolonged about 30°, strikes the star Ras Alhague, the brightest star in the constellation and the head of Ophiuchus. It is at one angle of an isosceles triangle, of which Altair is at the apex, and Vega the third angle.

Two constellations are here combined. Ophiuchus is represented as an old man, holding in his hands a writhing serpent.

Ras Algethi, marking the head of Hercules, lies just west of Ras Alhague.

Equally distant southeast and southwest of Ras Alhague are to be seen two stars close together, representing the shoulders of Ophiuchus. His foot rests on the Scorpion just above Antares.

The head of Serpens is the star group in the form of an "X" just below the Crown.

1604 indicates the spot where in that year a famous temporary star appeared, called Kepler's star.

Note the asterism the "Bull of Poniatowski" just east of γ. The star marked 70 is one of the most distant stars for which a parallax has been obtained. Its distance from the earth = 1,300,000 radii of the earth's orbit, or 120 quadrillion miles.

There is something remarkable in the central position of this gigantic figure. It is situated almost exactly in the mid-heavens, being nearly equidistant from the poles, and midway between the vernal and autumnal equinoxes.

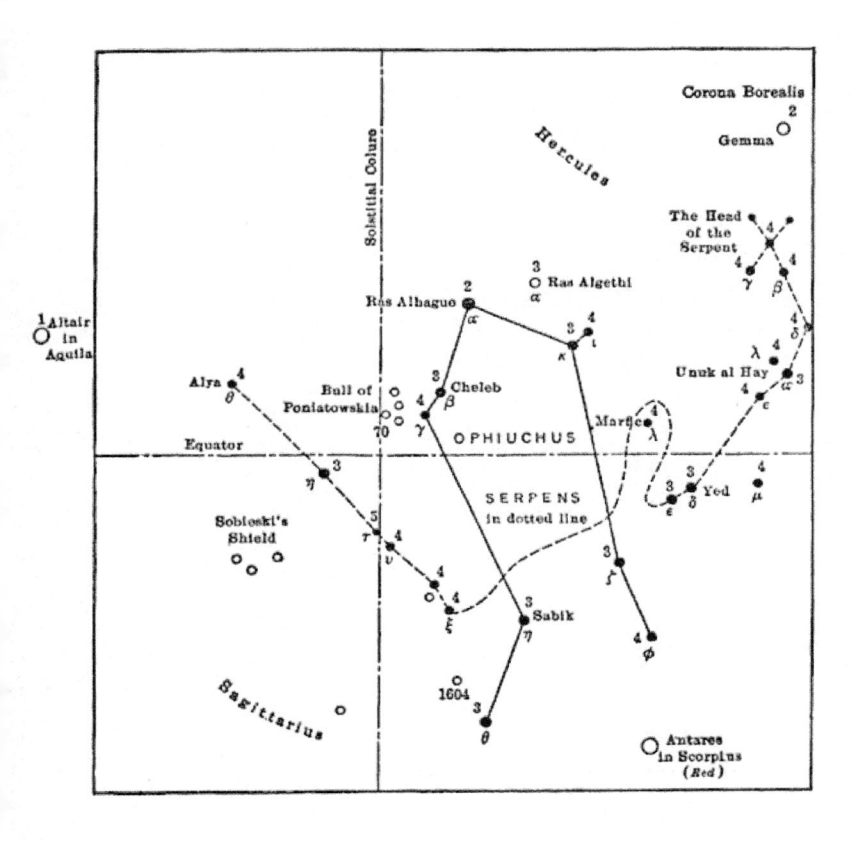

OPHIUCHUS & SERPENS

SCORPIUS (skôr'-pi-us)—THE SCORPION. (Face South.)

LOCATION.—Scorpius, one of the signs of the zodiac, is a beautiful star group, and one that is easily traced out. It lies just under the Serpent Bearer, between Sagittarius and Libra.

The resemblance to a Scorpion is not difficult to see, hence this constellation is perhaps the most aptly named of any.

The ruddy star Antares, the brightest star in the constellation, is in the heart of the Scorpion. It lies about 40° southwest of Ras Alhague, in Ophiuchus, and a little over 20° west of the bow of Sagittarius. The fact that it is the most brilliant star in this region of the sky renders its identity unmistakable. It is one of the reddest stars in the firmament.

There are several star clusters and double stars to be seen in this constellation. Their position is indicated in the diagram.

The curved tail of the Scorpion is very conspicuous. λ and υ are a striking pair and the fine clusters above them can be seen with the naked eye.

A record of a lunar occultation of β Scorpii in 295 B.C. is extant.

Note a pair just below β. They are known as ω^1 and ω^2.

In this region of the sky have appeared many of the brilliant temporary stars, the first one in astronomical annals being discovered in 134 B.C.

Scorpius is mentioned by all the early writers on astronomy and is supposed to be so named because in Egypt it was a sickly time of the year when the sun entered this sign.

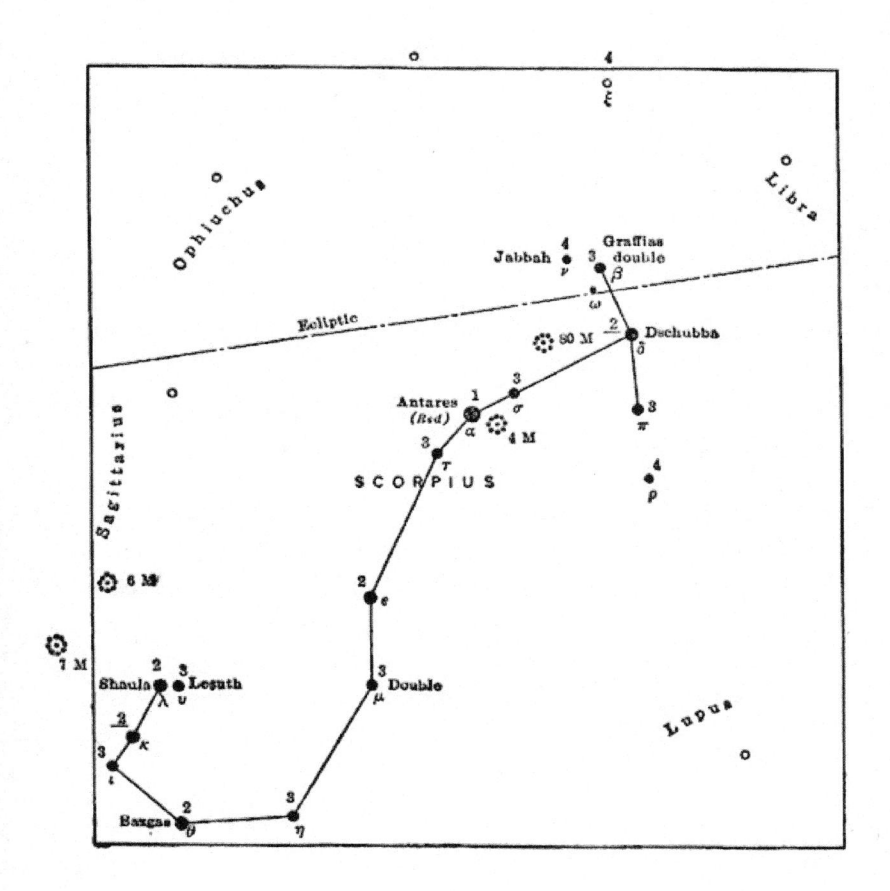

SCORPIUS

LIBRA (lī′-bra)—THE SCALES. (Face Southwest.)

LOCATION.—Libra is one of the signs of the zodiac, and lies between Virgo and Scorpius. Its two chief stars, α and β, may be recognized west of and above the head of the Scorpion.

The star ι Libræ is about 20° northwest of Antares in the Scorpion. Spica in Virgo, a star of the first magnitude, is a little over 20° northwest of α Libræ.

A quadrilateral is formed by the stars α, β, γ, ε, which characterizes the constellation.

The star α Libræ looks elongated. An opera-glass shows that it has a fifth-magnitude companion.

β is a pale green star. Its color is very unusual.

Lyra, Corona, and Hercules are almost directly overhead in the early evening, during July and August, and can best be observed in a reclining position. Thus placed, with an opera-glass to assist the vision, you may study to the best advantage the wonderful sight spread out before you, and search depths only measured by the power of your glass.

When the sun enters the sign Libra the days and nights are equal all over the world and seem to observe a certain equilibrium like a balance, hence the name of the constellation.

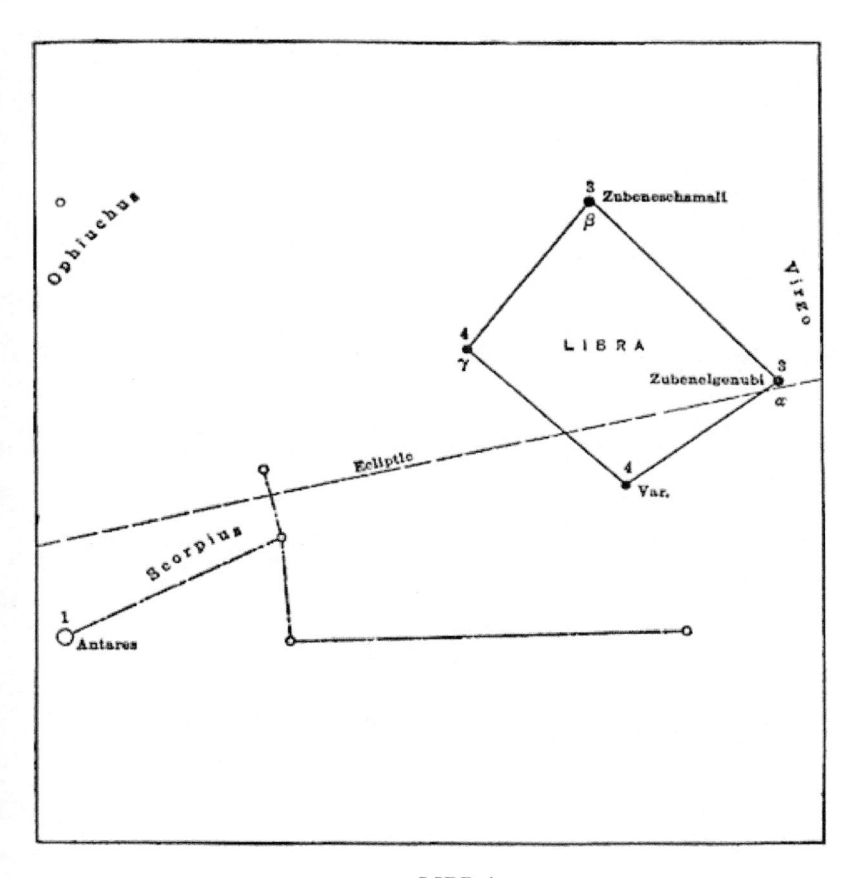

LIBRA

CORONA BOREALIS (kō-rō′nä bō-rē-a′-lis)—THE NORTHERN CROWN.

LOCATION.—A line drawn from α Cygni, to α Lyræ, and projected a little over 40°, terminates in the Crown, which lies between Hercules and Boötes, and just above the diamond-shaped group of stars in the head of the Serpent.

The characteristic semicircle resembling a crown is easily traced out. The principal stars are of the fourth magnitude excepting Gemma, which is a second-magnitude star and known as the "Pearl of the Crown."

Gemma, sometimes called Alphacca, forms with the stars Seginus and Arcturus, in Boötes, an isosceles triangle, the vertex of which is at Arcturus.

Close to ε a famous temporary appeared suddenly May 12, 1866, as a second-magnitude star. It was known as the "Blaze Star" and was visible to the naked eye only eight days, fading at that time to a tenth-magnitude star, and then rising to an eighth-magnitude, where it still remains.

The native Australians called this constellation "The Boomerang." To the Hebrews it was "Ataroth" and by this name it is known in the East to-day. No two of the seven stars composing the Crown are moving in the same direction or at the same rate.

α Coronæ is seventy-eight light years distant and sixty times brighter than the sun.

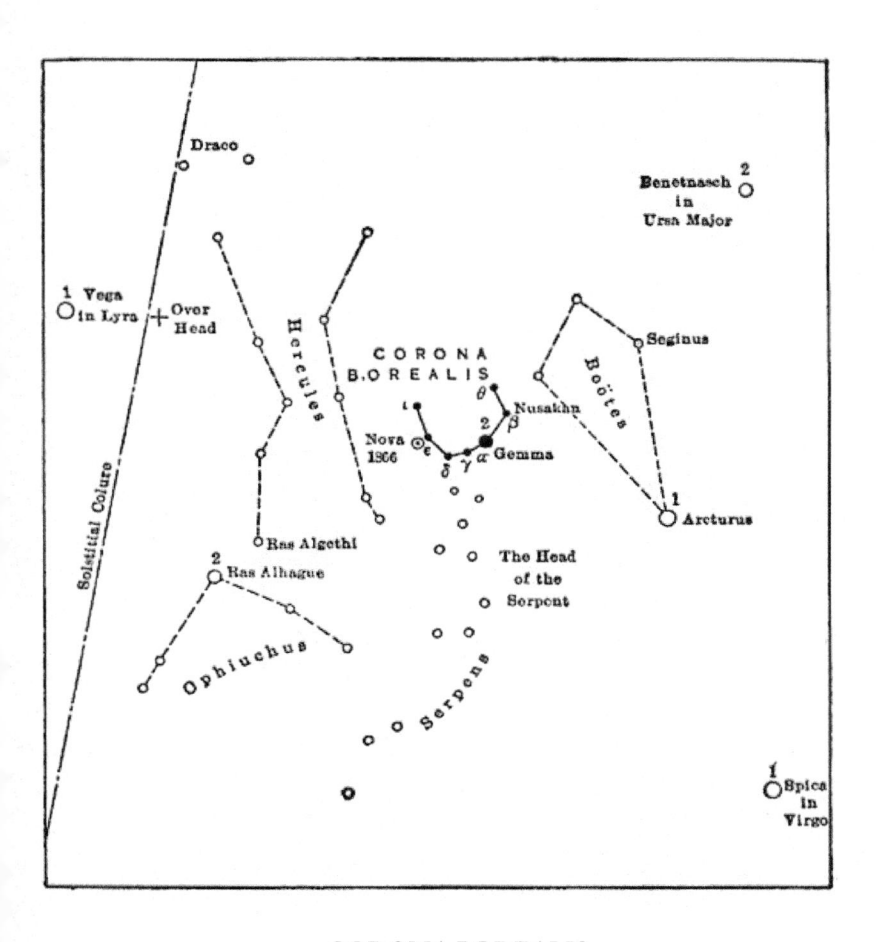

CORONA BOREALIS

HERCULES (her´-kū-lēz)—THE KNEELER.

LOCATION.—A line drawn from either Vega, in Lyra, or Altair, in Aquila, to Gemma, in Corona Borealis, passes through this constellation. The left foot of Hercules rests on the head of Draco, on the north, and his head nearly touches the head of Ophiuchus on the south.

The star in the head of Hercules, Ras Algethi, is about 25° southeast of Corona Borealis.

α Ophiuchi and α Herculis are only about 5° apart.

The cluster 13 M., the Halley Nebula, can be easily seen in an opera-glass. In a recent photograph of this cluster 50,000 stars are shown in an area of sky which would be entirely covered by the full moon.

Hercules occupies the part of the heavens toward which the sun is bearing the earth and planets at the rate of twelve miles a second or 373 million miles a year.

On a clear night the asterism Cerberus, the three-headed dog, which Hercules holds in his hand, can be seen.

This constellation is said to have been an object of worship in Phœnicia. There is a good deal of mystery about its origin. The ancient Greeks called it "The Phantom" and "The Man upon his Knees."

The stars ε, ζ, η, and π form a keystone shaped figure that serves to identify the constellation.

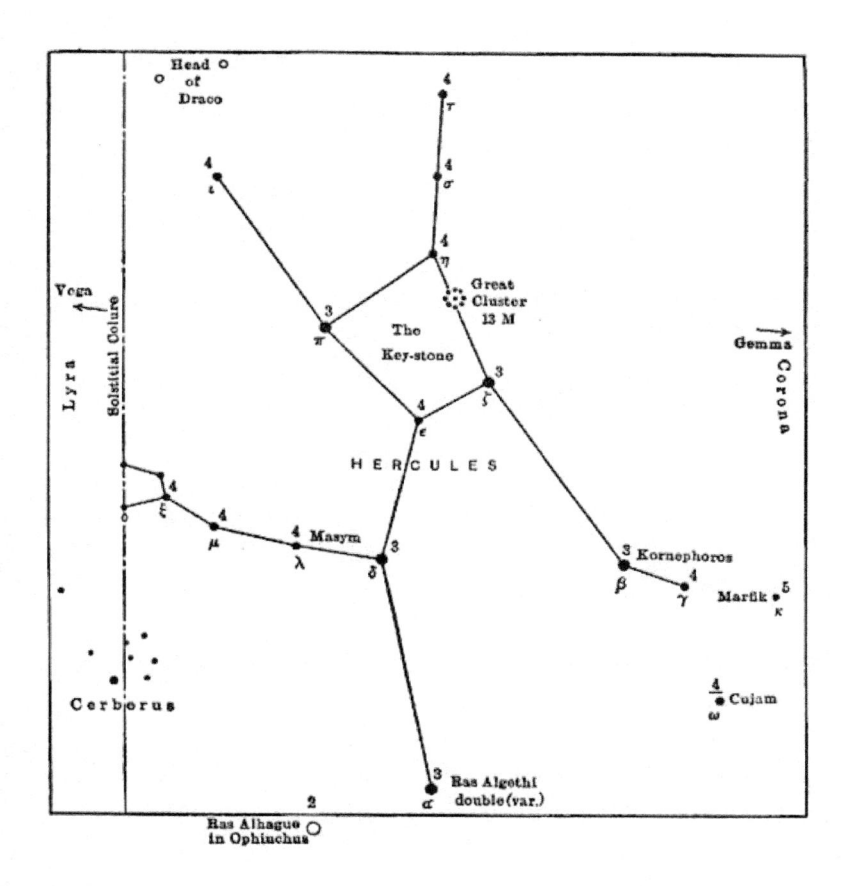

HERCULES

BOÖTES (bō-ō′tēz)—THE HERDSMAN, OR BEAR DRIVER. (Face West.)

LOCATION.—Boötes lies just west of the Crown, and east of Cor Caroli. It may be easily distinguished by the position and splendor of its principal star, Arcturus, which shines with a golden yellow lustre. It is about 35° east of Denebola, in Leo, and nearly as far north of Spica, in Virgo, and forms with these two a large equilateral triangle. A line drawn from ζ to η Ursæ Majoris and prolonged about 30° locates it, as does one from δ Herculis to γ Coronæ prolonged its length.

The brightest stars in Boötes outline a characteristic kite-shaped figure. Arcturus is mentioned in the Book of Job and is often referred to as "The Star of Job."

Three stars of the fourth magnitude are situated in the right hand. They are about 5° north of η Ursæ Majoris.

Contrast the color of Arcturus with Spica, Antares, and Vega.

The trapezium β, γ, δ, and μ, was called "The Female Wolves," by the Arabians; θ, ι, κ and λ, "The Whelps of the Hyenas." They knew the constellation as "The Vociferator."

Arcturus is the fourth brightest star in the northern hemisphere. It is 1000 times the size of our sun and rushes through space toward Virgo at the astounding rate of ninety miles a second. It is forty light years distant.

The ancient Greeks called this constellation "Lycaon," a name which signifies a Wolf. The Hebrew name for it was "The Barking Dog."

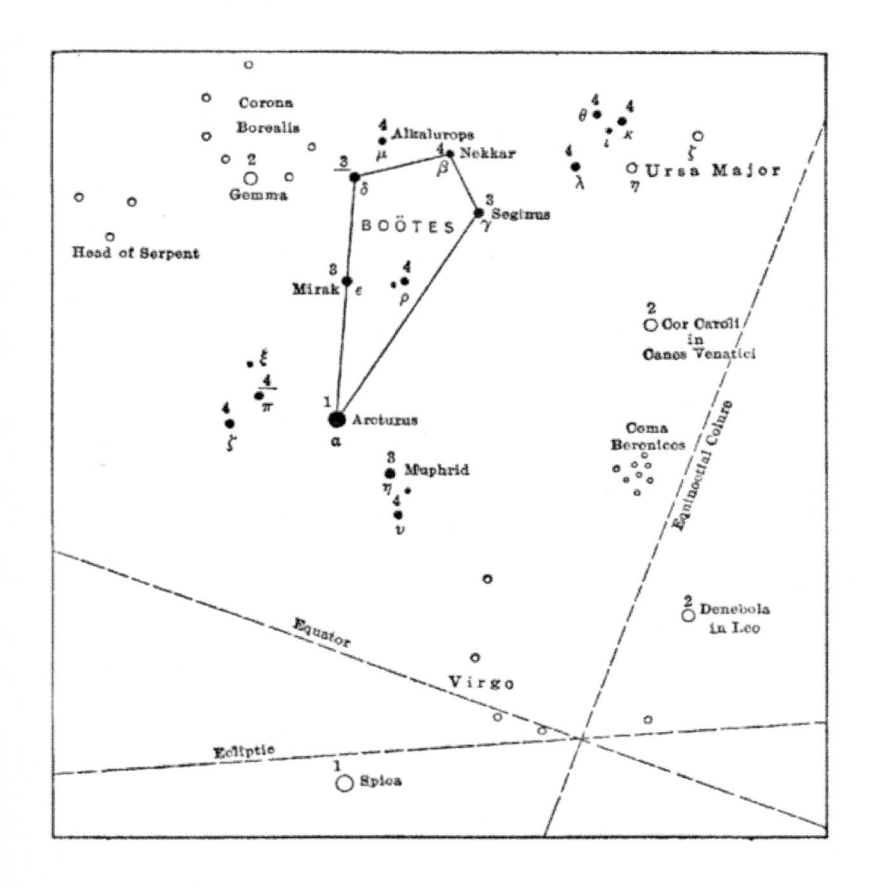

BOÖTES

VIRGO (ver'-gō)—THE VIRGIN. (Face West.)

LOCATION.—An imaginary line drawn from Antares in Scorpius through α Libræ and prolonged a little over 20° strikes Spica, the brightest star in Virgo, which star is about 30° southwest of Arcturus.

Arcturus, Cor Caroli, Denebola, and Spica form a figure about 50° in length, called the Diamond of Virgo.

The equator, ecliptic, and equinoctial colure intersect each other at a point close to the star η. This is called the autumnal equinox.

The star ε is known as the "Grape Gatherer." It is observed to rise just before the sun at vintage time.

Within the rude square formed by Denebola, and ε, γ, and β, Virginis, the telescope reveals many wonderful nebulæ; hence this region of the sky has been called "The Field of the Nebula."

Spica is an extremely beautiful pure white star. It rises a very little south of the exact eastern point on the horizon.

γ is a fine double star for a small telescope.

Virgo is mentioned by the astronomers of all ages. By the Egyptians it was intended to represent the goddess Isis, and the Greeks knew it as Ceres. Spica represents the ear of corn held in the Virgin's left hand.

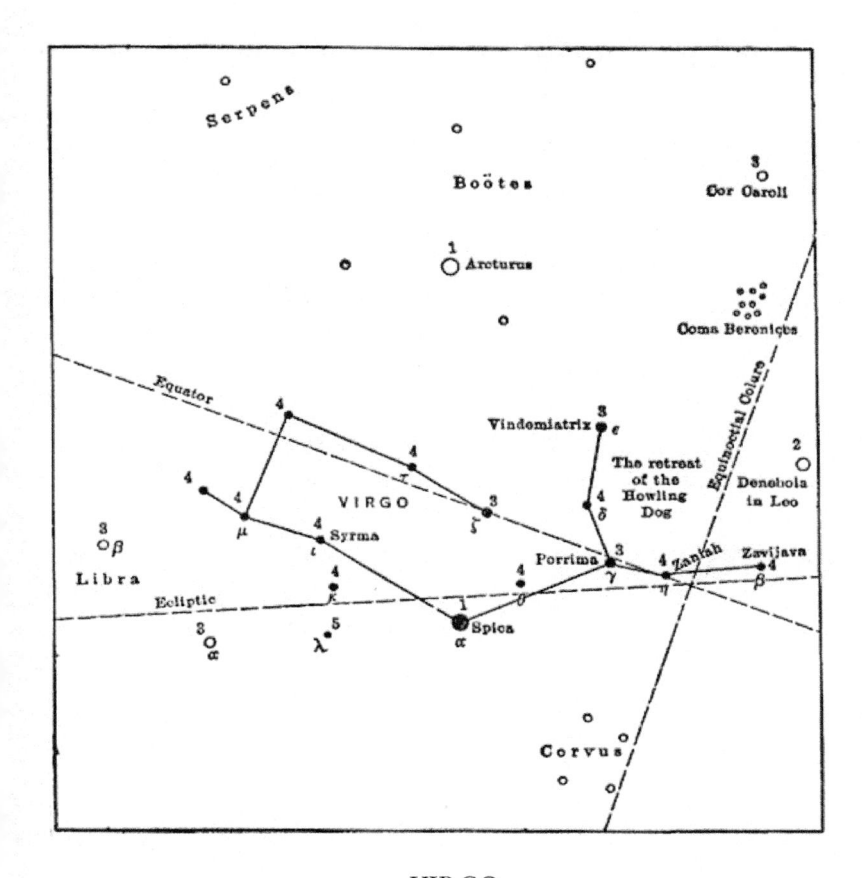

VIRGO

CANES VENATICI (kā'-nēz ve-nat'-i-cī)—THE HUNTING DOGS. (Face Northwest.)

LOCATION.—Cor Caroli, the bright star in this constellation, when on the meridian is about 17° south of ε Ursæ Majoris. A line drawn from η Ursæ Majoris, through Berenice's Hair, to Denebola, in Leo, passes through it. The dogs, Asterion and Chara, are represented as being held in leash by Boötes, the herdsman, in his pursuit of the Great Bear. Cor Caroli is in the southern hound, Chara, and represents the heart of Charles II of England. It is a beautiful double star in a small telescope.

The so-called "Diamond of Virgo," is clearly shown on this plate. It is formed by connecting with lines the stars Cor Caroli, Denebola, Spica, and Arcturus.

The fifth-magnitude star La Superba, about 7° north and 2½° west of Cor Caroli, is especially noteworthy because of the flashing brilliancy of its prismatic rays.

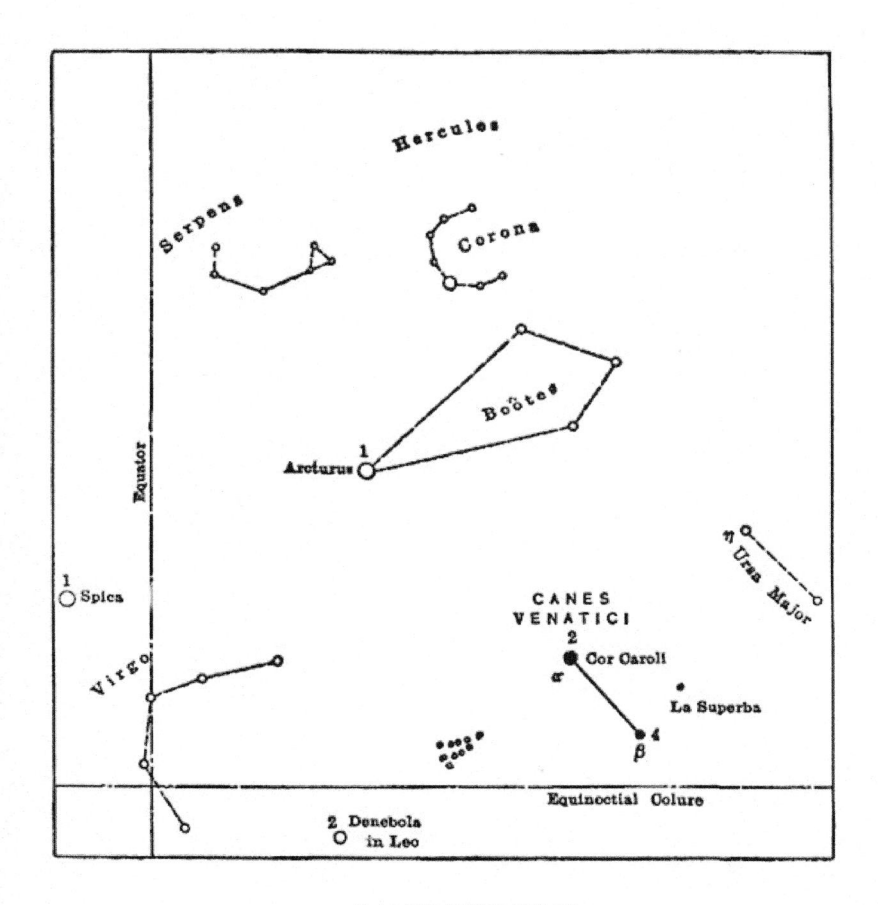

CANES VENATICI

METEORIC SHOWERS.

JULY TO OCTOBER.

Name of Shower	Date	Radiant Point	Characteristics	Other Dates of Observation	Location
Vulpeculids or Eta Sagittids	July 4	Between Cygnus and Delphinus	Sw.	June 13-July 7 Apr. 20, May 30	E.
Cygnids	July 19	Near Deneb (α) Cygni	Sh. Sw. F.	July 11-19, Aug. 22, July 6-Aug. 16	E.
(α)-(β)Perseids	July 25	Between (α) and (β) Persei	Sw. B. Sk. after 10 P.M.	July 23-Aug. 4 Sept. 15, Nov. 13	N.E.
Aquarids, a Conspicuous Shower	July 28	Near the water jar of Aquarius	Sl. B.		E.
Perseids, fine Shower	Aug. 10	Near (α) Persei	v. Sw. Sk.		N.E.
Kappa Cygnids	Aug. 17	Near the Dragon's head	Sw. B.T. Sh.	Jan. 17, Aug. 4, Aug. 21-25	S.E.
Alpha Aurigids	Aug. 21	Near Capella (α) Aurigæ	After 9.30 P.M. v. Sw. Sk.	Sept. 22, Oct. 2	N.E.
Omicron Draconids. Rich shower in 1879	Aug. 22	Near the Dragon's head	Sl. T.	Aug. 21-25	N

Name of Shower	Date	Radiant Point	Characteristics	Other Dates of Observation	Location
Epsilon Perseids	Sept. 7	Between Capella and the Pleiadesv.	After 10 P.M. Sw. Sk.	Aug. 21, 25, Sept. 6-8, 21, Nov. 29.	N.E.
Alpha Arietids	Sept. 21	Near Hamal (α) Arietis	Sl. T.	Aug. 12, Oct. 7	E.
Gamma Pegasids	Sept. 22	Near and S.E. of Great Sq.	Sl.	July 31, Aug. 5, etc.	E.

The Perseids are of a yellowish color, and move with medium velocity. Their line of flight is from northeast to southwest. They are probably visible for more than a month, from the latter half of July to the last week in August.

The August meteors are known as the "Tears of St. Lawrence."

The Abbreviations under *Characteristics* are as follows:

v.—very
Sl.—Slow
Sk.—Streak-leaving meteors.
M.—Moderately
B.—Bright
T.—Train-leaving meteors.
Sw.—Swift
F.—Faint
Sh.—Short meteors.

THE CONSTELLATIONS OF AUTUMN.

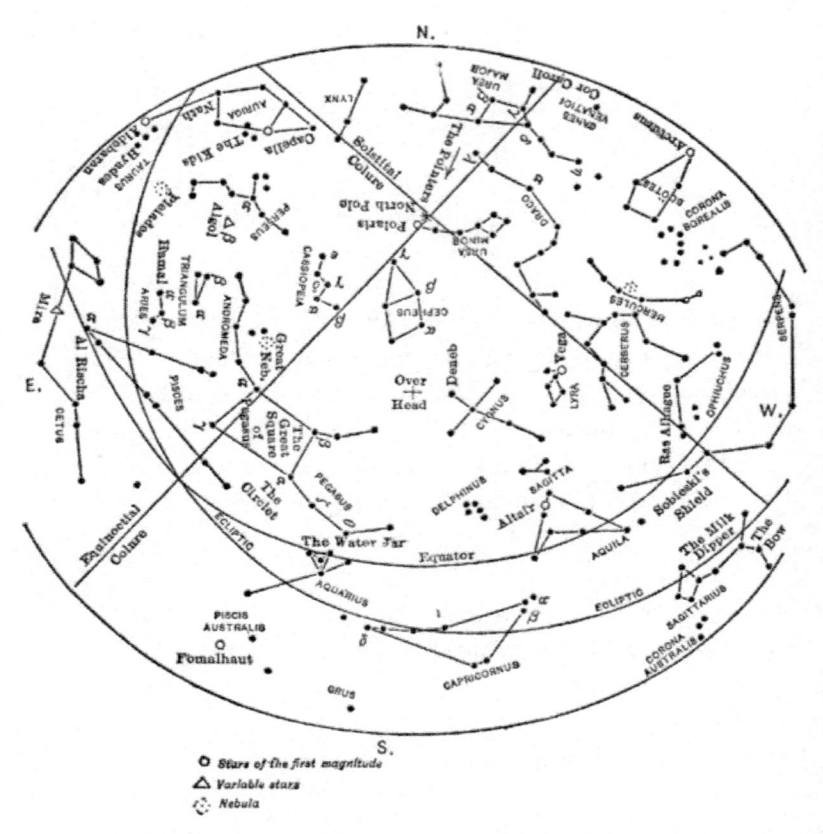

Map showing the principal stars visible from Lat. 40° N. at 9 o'clock, October first.

CASSIOPEIA (kas-i-ō-pē´-ya)—THE LADY IN THE CHAIR.
(Face North.)

LOCATION.—A line drawn from δ Ursæ Majoris, through Polaris, strikes α Cassiopeiæ. It is situated the same distance from Polaris as Ursa Major, and about midway between Polaris and the zenith in the Milky Way. Cassiopeia is characterized by a zigzag row of stars which form a rude "W," but in mid-autumn, to an observer facing north, the "W" appears more like an "M," and is almost overhead. Note the spot marked 1572. This is where a very famous temporary star appeared in that year. It was bright enough at one time to be seen in full sunshine. The star η is sixteen light years distant. Caph is equidistant from the Pole, and exactly opposite the star Megres in Ursa Major; with α Andromedæ and γ Pegasi it marks the equinoctial colure. These stars are known as "The Three Guides."

The chair can be readily traced out; β, α, and γ mark three of the four corners of the back, and δ and ϵ, one of the front legs. The word "Bagdei," made up of the letters for the principal stars, assists the memory.

The stars γ and β are pointer stars to a fifth-magnitude star the lucida of the asterism Lacerta, the lizard about 15° from β.

Cassiopeia makes an excellent illuminated clock. When β is above Polaris it is noon, when it is in the west at right angles to its first position it is 6 P.M. At midnight it is on the northern horizon, and at 6 P.M. it is due east.

This is sidereal time which agrees with mean time on March 22d, and gains on the latter at the rate of two hours a month.

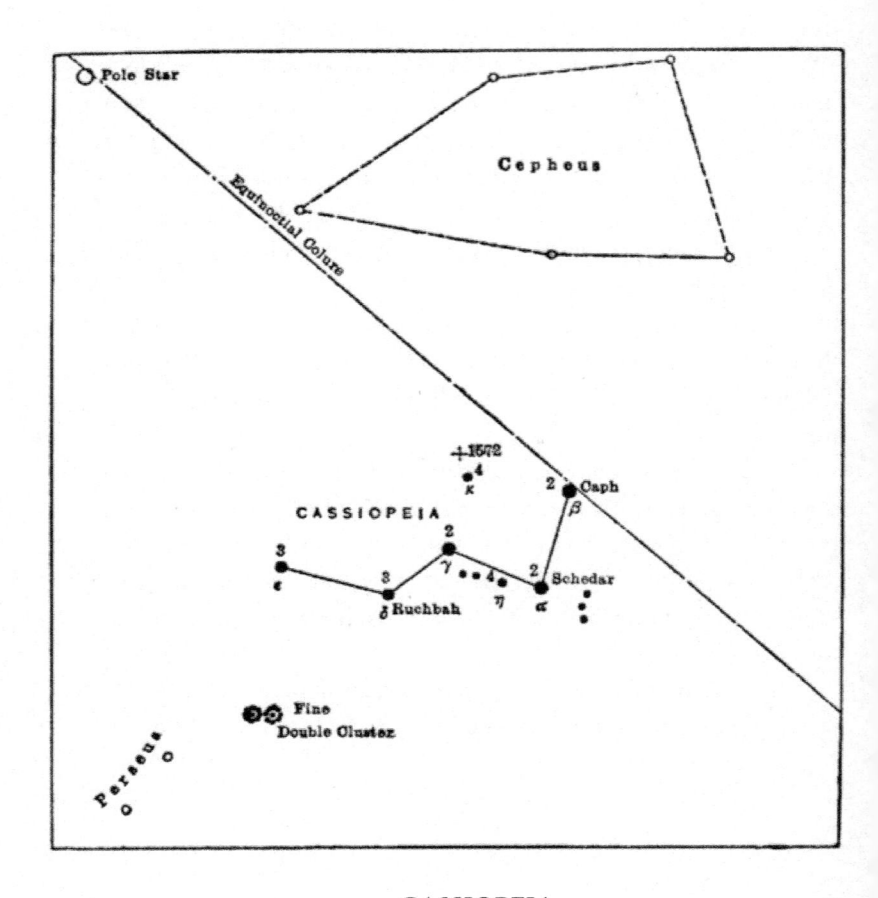

CASSIOPEIA

CEPHEUS (sē′-fūs) (Face North.)

LOCATION.—A line drawn from α to β Cassiopeiæ and prolonged about 18° strikes α Cephei. The nearest bright star west of Polaris is γ Cephei. Cepheus is an inconspicuous constellation, lying partly in the Milky Way. A view of this constellation through an opera-glass will repay the observer. Cepheus is characterized by a rude square, one side of which is the base of an isosceles triangle. Look for the so-called garnet star μ, probably the reddest star visible to the naked eye in the United States. The star ζ has a blue companion star.

α forms an equilateral triangle with Polaris and ε Cassiopeiæ.

It is claimed that Cepheus was known to the Chaldæans twenty-three centuries before our era.

Surrounding δ, ε, ζ, and λ, which mark the king's head, is a vacant space in the Milky Way, similar to the Coal Sack of Cygnus.

About 4° from γ, in the direction of κ is a pretty pair of sixth-magnitude stars.

Owing to precession, γ, β, and α Cephei will be successively the Pole Star in 4500, 6000, and 7500 A.D. respectively.

δ is a double whose components are yellow and blue. It is an interesting variable changing from magnitude 3.7 to 4.9 at intervals of 5 days 8 hours 47 minutes. As it is three times as bright at maximum as at minimum and can be observed with the naked eye its variations are well worth observing.

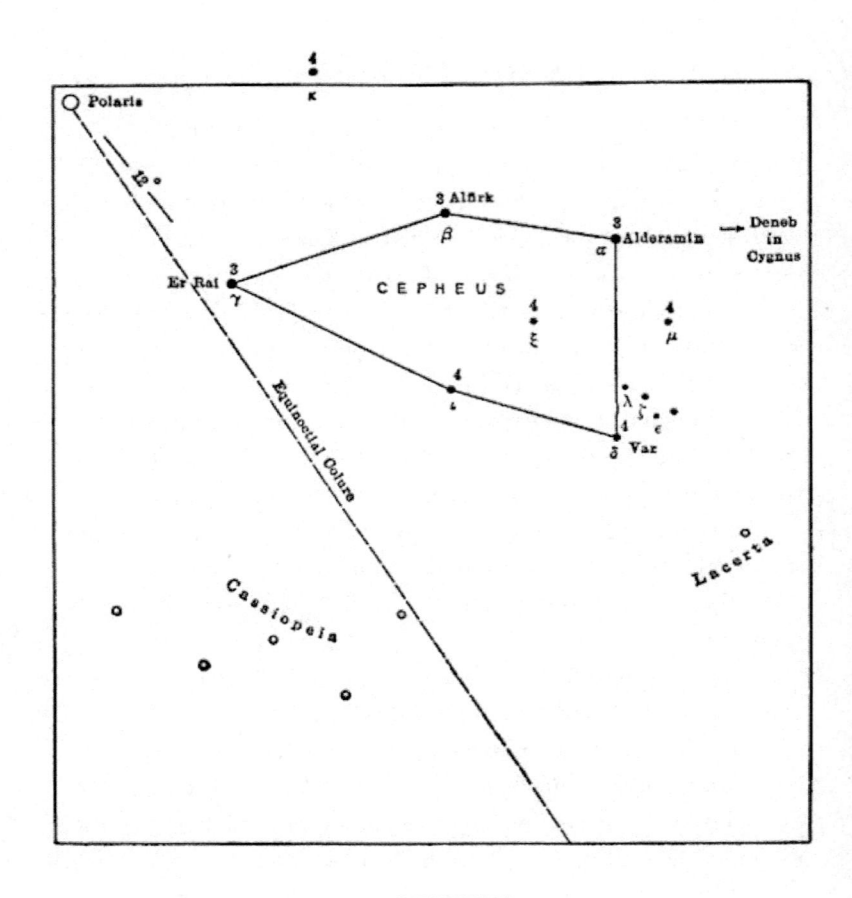

CEPHEUS

PEGASUS (peg′-a-sus)—THE WINGED HORSE. (Face South.)

LOCATION.—One corner of the Great Square is found by drawing a line from Polaris to Cassiopeia, and prolonging it an equal distance.

The Great Square is a stellar landmark. Three of the corners of the square are marked by stars in Pegasus; the fourth, and northeastern, corner is marked by the star Alpheratz in Andromeda. Each side of the square is about 18° long.

The horse is generally seen upside down, with his fore feet projected up into the sky. Only the head, neck, and fore feet are represented. The star Enif marks the nose.

π is an interesting double, easily seen in an opera-glass. All the stars of the Square are approaching us at an inconceivable speed.

The position of the asterism Equus or Equūleus, the Little Horse, or Horse's Head, is shown in the diagram.

Delphinus, the water jar of Aquarius, and the circlet in the Western Fish, are all in the vicinity of Pegasus, and indicated in the diagram.

The winged horse is found on coins of Corinth 500 to 430 B.C. The Greeks called this constellation ίππoσ.

Pegasus seems to have been regarded in Phœnicia and Egypt as the sky emblem of a ship.

Within the area of the Square Argelander counted thirty naked-eye stars.

Note a fine pair in Equūleus just west of the star Enif in Pegasus.

The position of the equinoctial colure is defined by a line connecting Polaris, β Cassiopeiæ, α Andromedæ, and γ Pegasi.

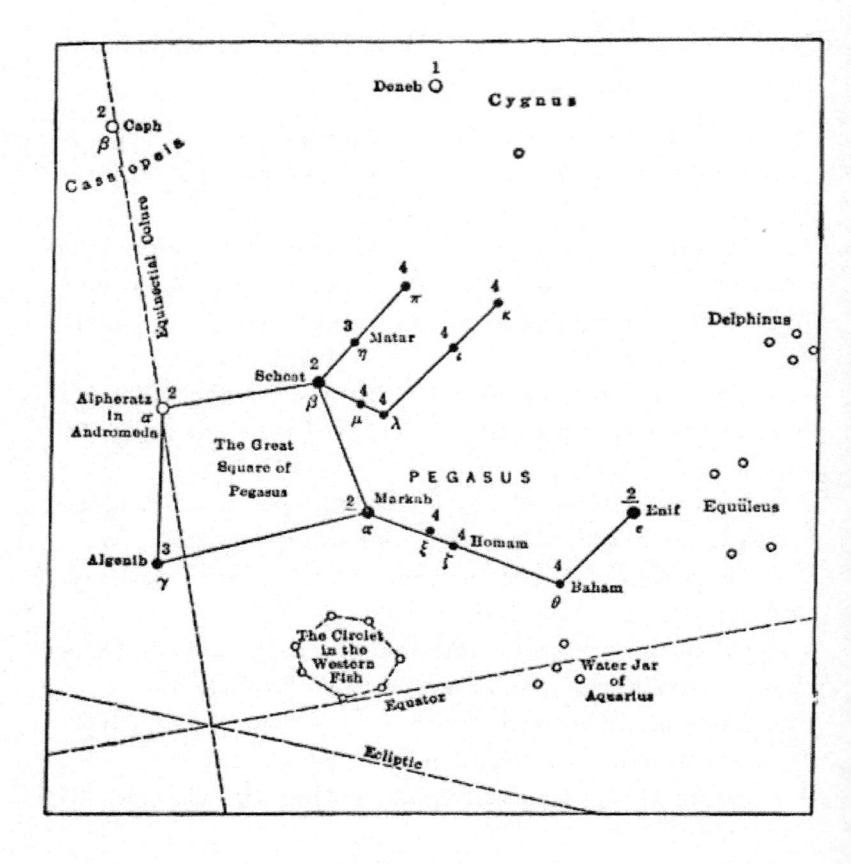

PEGASUS

ANDROMEDA (an-drom'-e-dä)—THE CHAINED LADY.

LOCATION.—The star α Alpheratz is at the northeastern corner of the great square of Pegasus, one of the stellar landmarks.

Running east from α, at almost equal distances, are four other stars, two of which are of the second magnitude. The most easterly one is β Persei, known as Algol, the famous variable. Lines connecting the stars γ Andromedæ, Algol, and α Persei form a right-angled triangle. The right angle is marked by Algol.

The chief object of interest in this constellation is the great nebula, the first to be discovered. It can be seen by the naked eye and it is a fine sight in an opera-glass. Its location is indicated in the diagram.

The star γ is the radiant point of the Bielid meteors, looked for in November. It is a colored double visible in a 3" glass.

The great nebula has been called the "Queen of the Nebulæ." It is said to have been known as far back as A.D. 905, and it was described 986 A.D. as the "Little Cloud."

Andromeda is very favorable for observation in September, low in the eastern sky.

Note the characteristic "Y" shaped asterism known as Gloria Frederika or Frederik's Glory. It lies about at the apex of a nearly isosceles triangle of which a line connecting Alpheratz and β Pegasi is the base. A line drawn from δ to α Cassiopeiæ and prolonged a little over twice its length points it out.

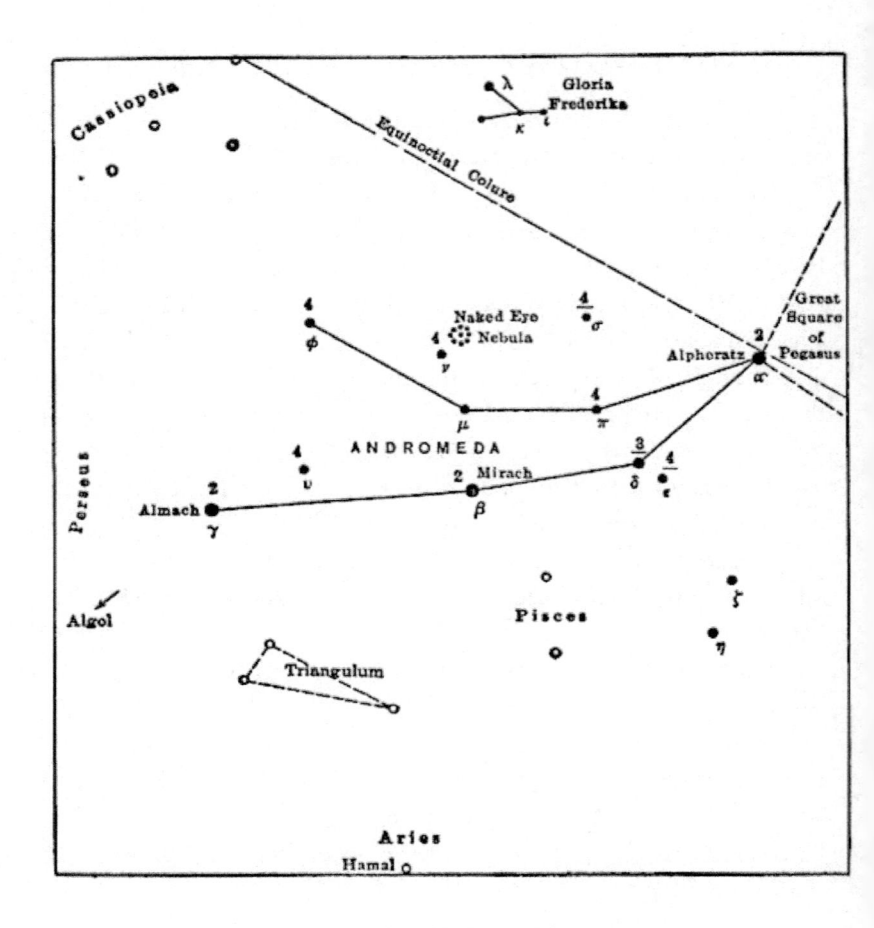

ANDROMEDA

PERSEUS (per´-sūs)—THE CHAMPION. (Face Northeast.)

LOCATION.—α Persei lies on a line drawn from β to γ Andromedæ, and is about 9° from the latter. The most striking feature in Perseus is the so-called "segment of Perseus," a curve of stars beginning about 12° below Cassiopeia, and curving toward Ursa Major. Note the famous variable Algol the Demon star. It represents the Medusa's head which Perseus holds in his hand. It varies from the second to the fourth magnitude in about three and one-half hours, and back again in the same time, after which it remains steadily brilliant for two and three-quarters days, when the same change recurs. Algenib and Algol form with γ Andromedæ, a right-angled triangle.

Note a dull red star near Algol, and a pretty pair just above Algenib.

An opera-glass reveals much that is worthy of observation in this region of the sky. It has been said of the clusters between Cassiopeia and Perseus that they form the most striking sidereal spectacle in the northern heavens. They are visible to the naked eye. Algenib never sets in the latitude of New York, just touching the horizon at its lower culmination. It is estimated that Algol is a little over a million miles in diameter, η has three faint stars on one side nearly in a line, and one on the other—a miniature representation of Jupiter and his satellites.

Algol, when on the meridian of New York City, is only one tenth of a degree from the zenith point. This remarkable variable has a dark companion star revolving near it obscuring its light in part from us at stated intervals. By means of the spectroscope the speed diameter and mass of this invisible star has been reckoned.

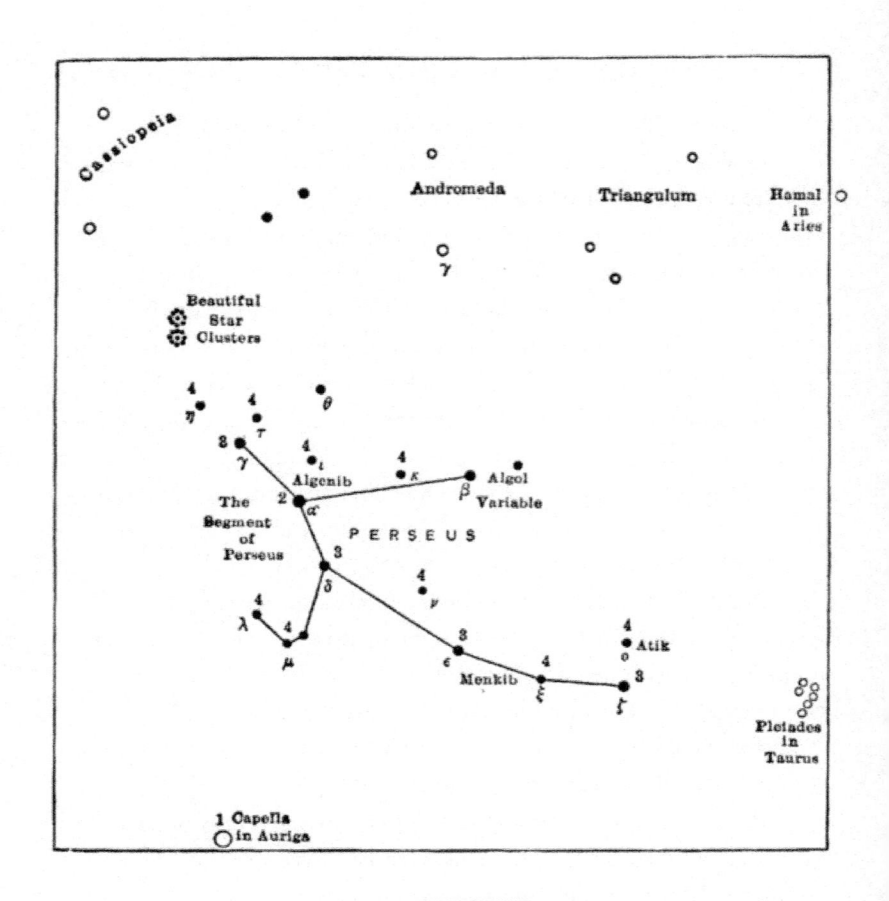

PERSEUS

PISCES (pis′ēz)—THE FISHES. (Face Southeast.)

LOCATION.—This constellation is represented by two fishes each with a ribbon tied to its tail. One, the Northern Fish, lies just below β Andromedæ,—the other, represented by the circlet, is just below Pegasus. The ribbons, represented by streams of faint stars, from a "V" with elongated sides, and terminate in the star Al Rischa, The Knot.

Below ω, and to the east of λ the spot marked (*) is the place which the sun occupies at the time of the equinox. It is one of the two crossing places of the equinoctial, or equator, of the heavens, and the ecliptic, or sun's path.

Below Pisces is Cetus, the Whale.

Pisces is thought to have taken its name from its coincidence with the sun during the rainy season.

Three distinct conjunctions of Jupiter and Saturn took place in this constellation in the year 747 of Rome.

Pisces was considered the national constellation of the Jews, as well as a tribal symbol.

In 1881, Jupiter, Saturn, and Venus were grouped together in Pisces.

The Circlet is a very striking group forming a pentagon. The glass reveals two faint stars in addition, making the figure seven-sided or elliptical in form.

As to the number of the stars as classified according to their magnitude, that is their brightness, it may be mentioned that there are approximately 20 stars of the first magnitude, 65 of the second, 300 of the third, and 450 of the fourth. We cannot see stars fainter than the sixth magnitude with the naked eye.

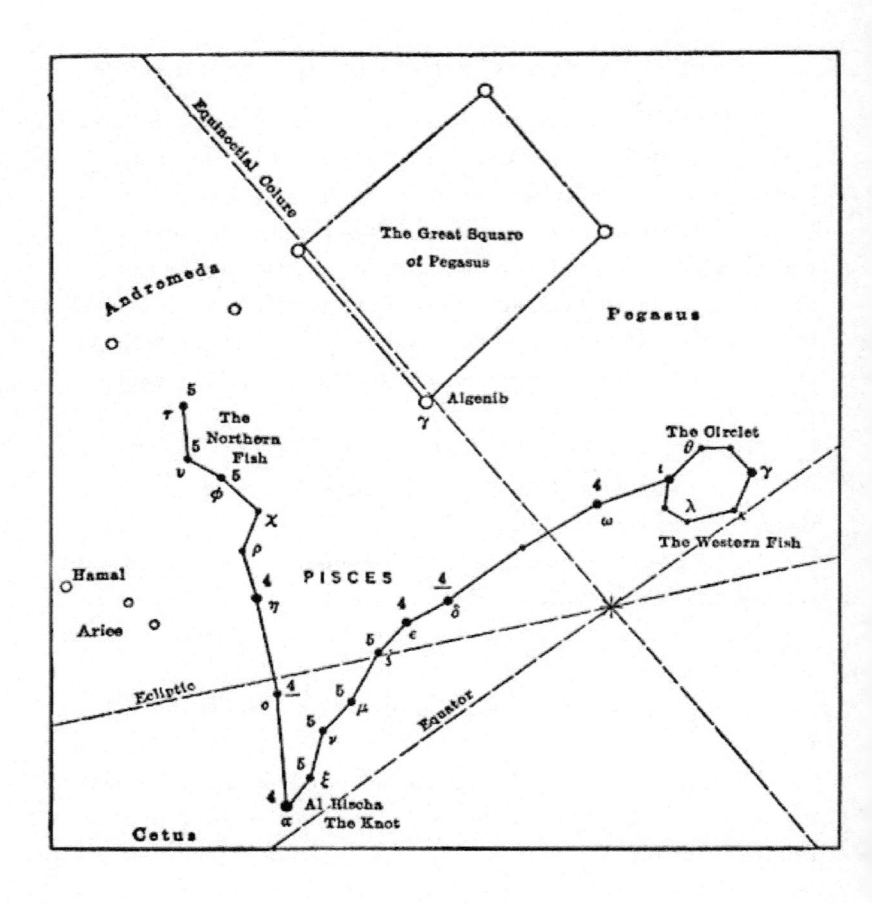

PISCES

TRIANGULUM (trī-an'-gū-lum)—THE TRIANGLE. (Face East.)

LOCATION.—A line drawn from the star γ Pegasi to Algol in Perseus passes through β Trianguli.

The triangle is clearly defined and a beautiful figure. It lies just below Andromeda, and above Aries.

Triangulum is a very ancient constellation, being formerly named Deltoton, from the Greek letter Delta Δ.

It was in this locality that Piazzi discovered the asteroid Ceres, January 1, 1800.

α Trianguli is sometimes called "Caput Trianguli."

α and β Trianguli were known as "The Scale Beam." According to Argelander the constellation contains fifteen stars.

The Triangle has been likened to the Trinity, and the Mitre of St. Peter.

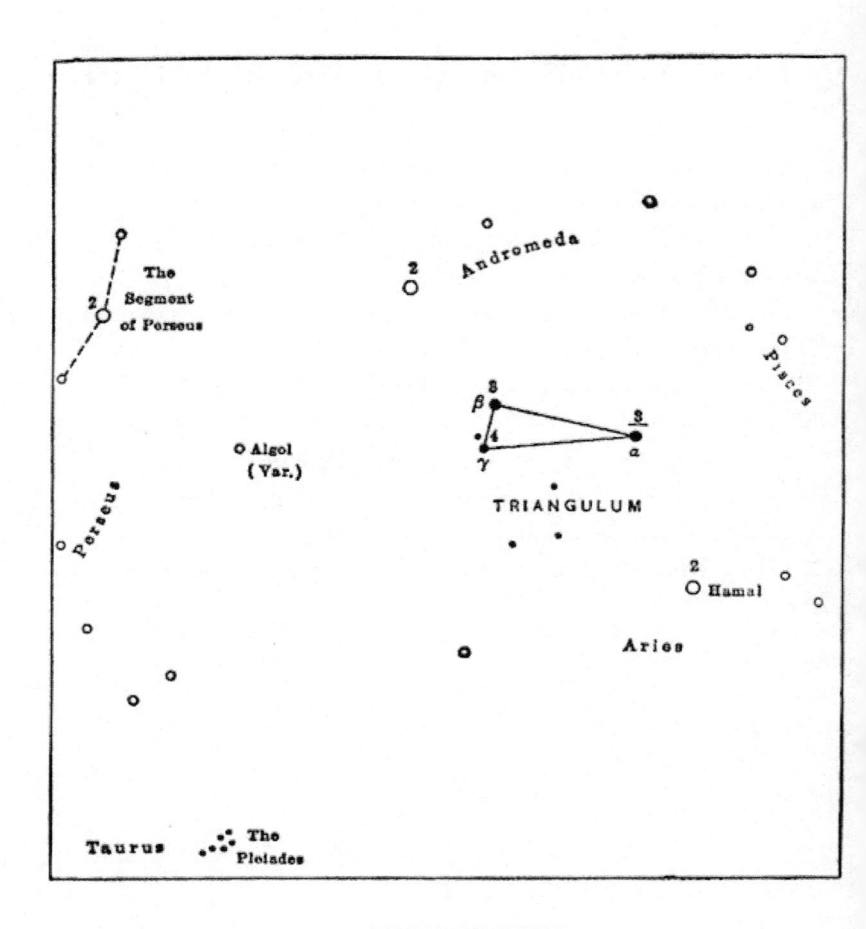

TRIANGULUM

AQUARIUS (a-kwā′ri-us)—THE WATER CARRIER. (Face Southwest.)

LOCATION.—A line drawn from β Pegasi to α of the same constellation, and prolonged as far again, ends just east of the so-called water jar of Aquarius, which is formed by a group of four stars in the form of a "Y," as indicated in the diagram. The Arabians called these four stars a tent.

The jar is represented as inverted, allowing a stream of water represented by dim stars in pairs and groups of three stars, to descend, ending in the bright star Fomalhaut, the mouth of the Southern Fish.

A rough map of South America can be traced in the stars θ, λ, τ, δ, 88, ι.

A rude dipper can be made out in the western part of the constellation, formed of the stars α, β, ν, ε.

The stars τ and ζ are doubles. Of the former pair, one is white, the other orange in color. Fomalhaut was the object of sunrise worship in the temple of Demeter at Eleusis in 500 B.C. The ancients called this region of the sky "the Sea."

In the vicinity of δ, Mayer observed in 1756 what he termed a fixed star. Herschel thought it a comet. It proved to be the planet Uranus.

ζ is almost exactly on the celestial equator.

λ is a red star, the most prominent of the first stars in the stream. The stars in Piscis Australis can be traced out with an opera-glass.

Fomalhaut and Capella, in Auriga, rise almost exactly at the same minute.

Fomalhaut is one of the four "royal stars" of astrology. The others are Regulus, Antares, and Aldebaran.

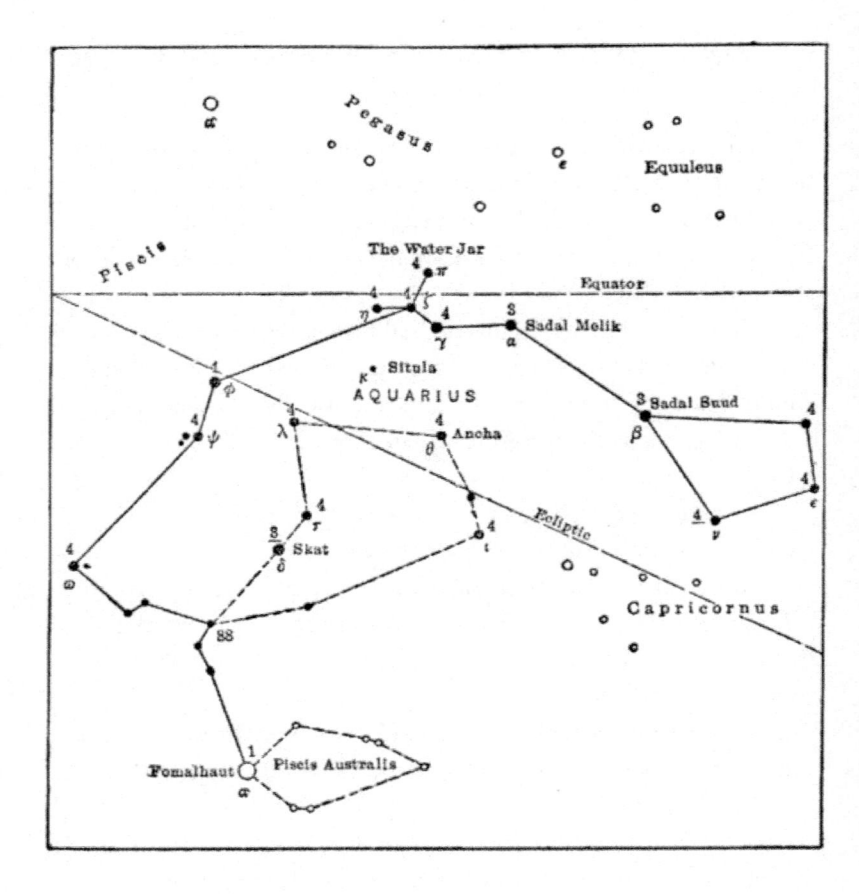

AQUARIUS

CAPRICORNUS (kap-ri-kôr'-nus)—THE SEA GOAT. (Face Southwest.)

LOCATION.—A line drawn from α Pegasi through ζ and θ in the same constellation, and projected about 25°, strikes α and β in Capricornus.

This constellation contains three principal stars—α and β mentioned above, and δ about 20° east of them.

The water jar of Aquarius is about the same distance northeast of δ Capricorni that Fomalhaut, in the Southern Fish, is southeast of it.

α has a companion which can be seen by the naked eye. It is a fine sight in an opera-glass. These two stars are gradually separating.

β is a double star, one being blue, the other yellow.

The constellation resembles a chapeau, or peaked hat, upside down.

The stars in the head of the Sea Goat, α and β are only 2° apart, and can hardly be mistaken by an observer facing the southwestern sky during the early evening in autumn.

Five degrees east of δ is the point announced by Le Verrier as the position of his predicted new planet, Neptune.

Flammarion claims that the Chinese astronomers noted the five planets in conjunction in Capricornus, in the year 2449 B.C.

The sign of the Goat was called by the ancient Orientalists "The Southern Gate of the Sun."

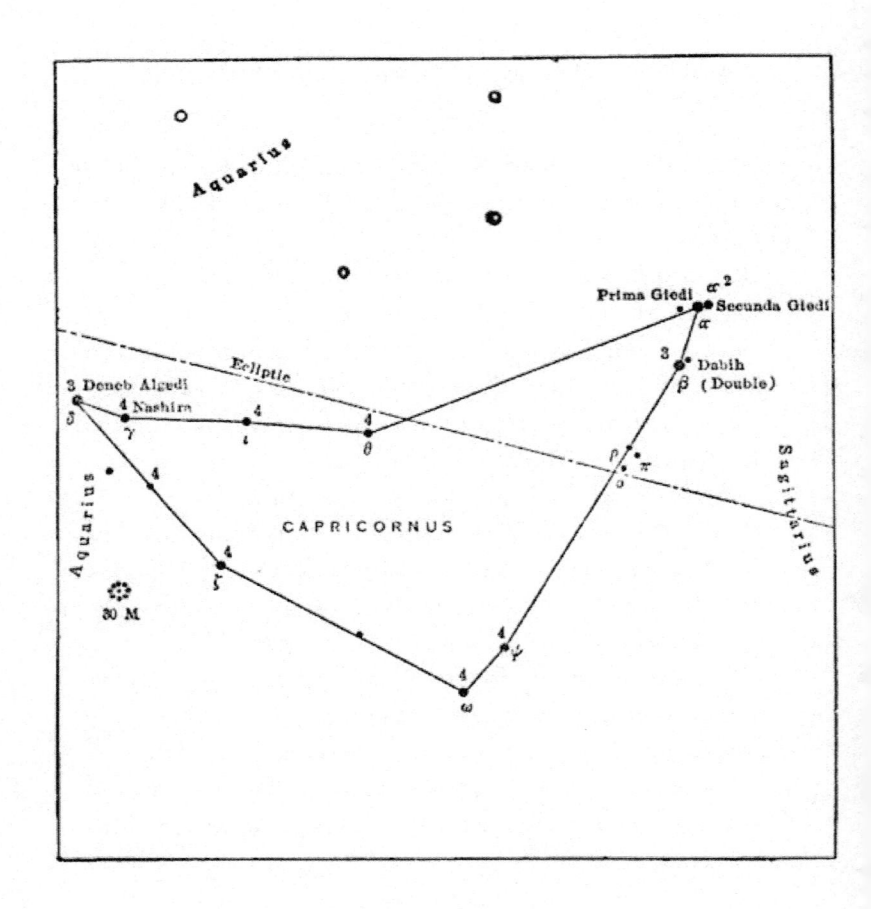

CAPRICORNUS

ARIES (ā'-ri-ēz)—THE RAM. (Face Southeast.)

LOCATION.—The star α in Aries, known as Hamal, and sometimes as Arietis, a star of the second magnitude, is about 7° south of α Trianguli. A line drawn from the Pole Star to γ Andromedæ, and prolonged about 20°, ends at Hamal.

Aries contains three principal stars, forming a characteristic obtuse-angled triangle.

The star γ Arietis was one of the first double stars discovered. A telescope is required to split it. Hamal lies near the path of the moon, and is one of the stars from which longitude is reckoned.

Below Aries may be seen the characteristic pentagon in the head of Cetus, the Whale.

More than two thousand years ago Aries was the leading constellation of the zodiac, and now stands first in the list of zodiacal signs.

The Arabians knew this constellation as Al Hamal, the sheep.

β and γ are one instance out of many where stars of more than ordinary brightness are seen together in pairs, the brightest star being generally on the east.

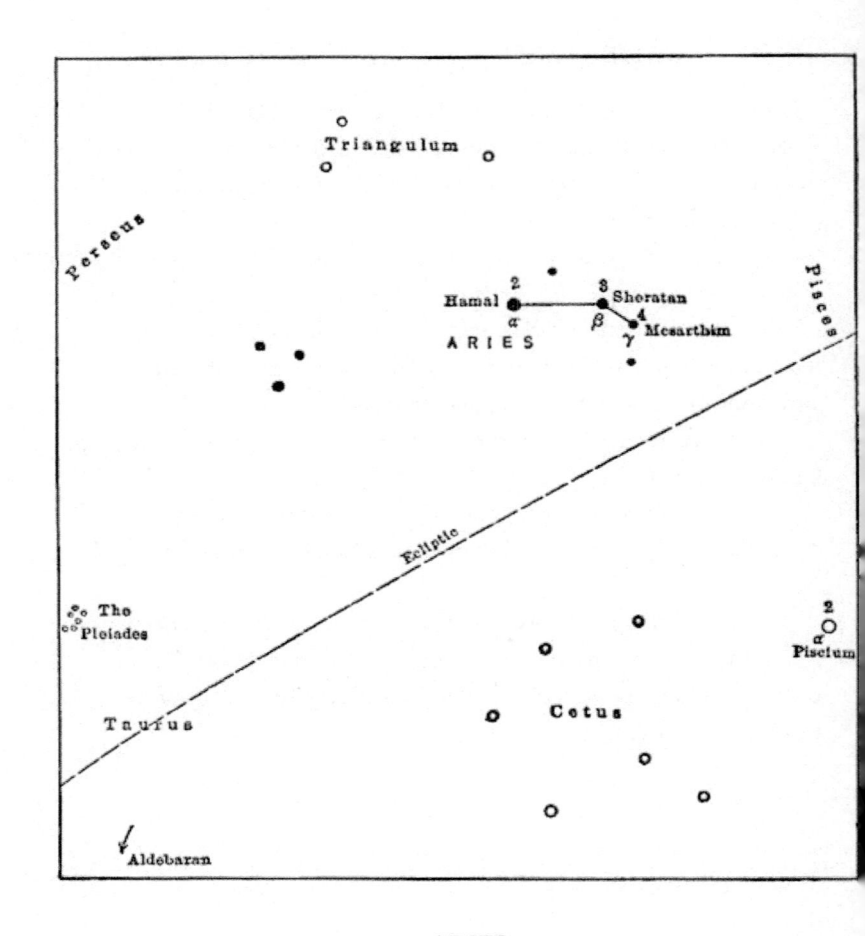

ARIES

CETUS (sē′-tus)—THE WHALE. (Face Southeast.)

LOCATION.—A line drawn from Polaris, to δ Cassiopeiæ, and prolonged two and one third times its original length, reaches the centre of this constellation.

It lies just below Aries and the Triangle, and resembles the figure of the prehistoric icthyosaurus, while some see in the outline an easy chair. The head of the beast is characterized by a clearly traced pentagon, about 20° southeast of Aries. The brightest star in the constellation is α of the second magnitude. It is at one apex of the pentagon, about 15° east of Al Rischa in Pisces, and 37° directly south of Algol.

The noted variable Mira also known as ο Ceti is the chief object of interest in this constellation.

It was discovered by Fabricius in 1596 and varies from the ninth magnitude to the third or fourth in a period of 334 days. It can be observed during its entire range with a 3″ glass.

In 1779 Mira is reported to have been as bright as the first-magnitude star Aldebaran. It lies almost exactly on a line joining γ and ζ Ceti a little nearer the former. Ten degrees south of it are four faint stars about 3° apart forming a square.

τ Ceti is one of our nearest neighbors at a distance of nine light years.

ζ is a naked-eye double star.

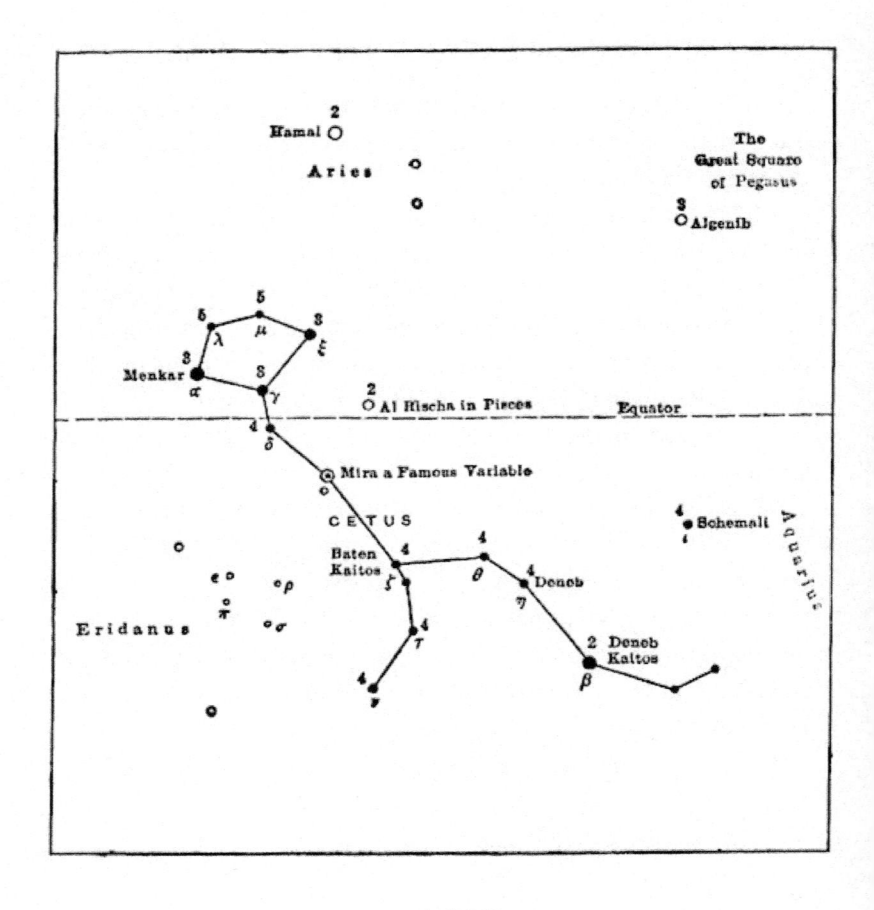

CETUS

MUSCA (mus´-kä)—THE FLY. (Face Southeast.)

LOCATION.—Musca lies between Triangulum and Aries, the diagram clearly defining its position.

The four stars composing it form a group shaped like the letter "Y." There is nothing of particular interest to be noted in this asterism. It does not appear on modern star charts and is considered obsolete.

So great is the distance that separates us from the stars that as for the great majority had they been blotted out of existence before the Christian era, we of to-day should still receive their light and seem to see them just as we do. When we scan the nocturnal skies we study ancient history. We do not see the stars as they are but as they were centuries on centuries ago.

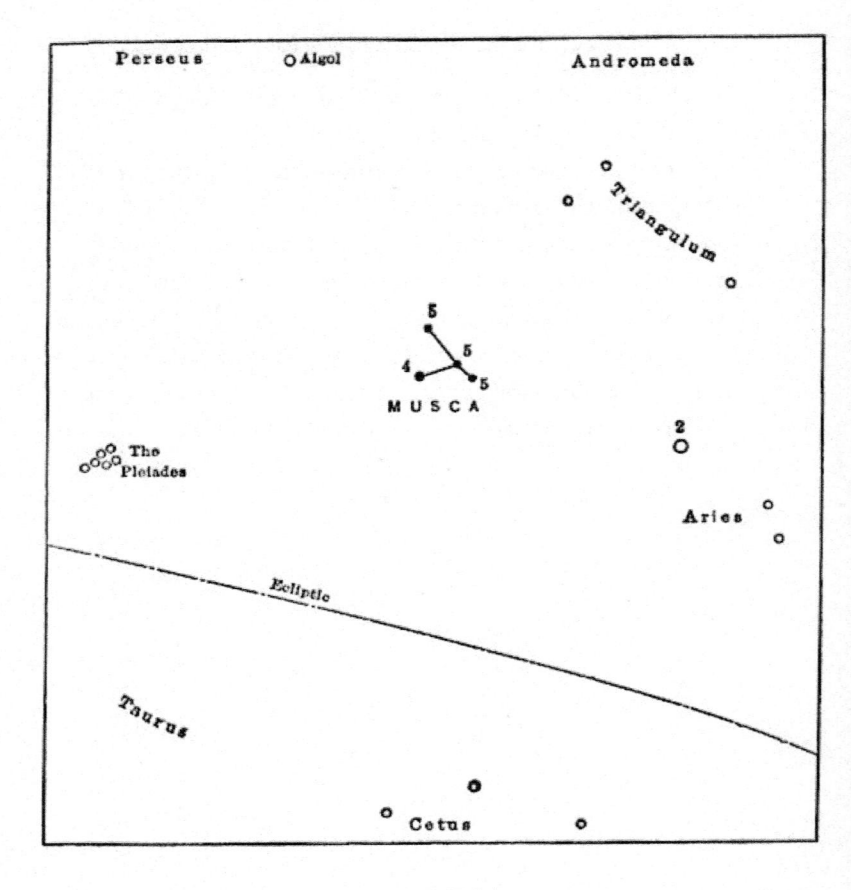

MUSCA

METEORIC SHOWERS.

OCTOBER TO JANUARY.

Name of Shower	Date	Radiant Point	Characteristics	Other Dates Of Observation	Location
Ursids	Oct. 4	Between Great Bear's head and Polaris	Sw. Sk.	Aug. 20-24	N.
Epsilon Arietids Rich shower 1877	Oct. 14	East of Hamal, near Musca	M. Sw.	Oct. 11-24, Oct. 30- Nov. 4	E.
Orionids Fine shower	Oct. 18	Near Alhena in Gemini	After 11 P.M. Sw. Sk.	Oct. 16-22	E.
Delta Geminids	Oct. 29	Near Castor and Pollux	After 10 P.M. v.Sw. Sk.	Nov. 7, Dec. 4, Oct. 16-22	N.E.
(e) Taurids. Rich shower in 1886	Nov. 2	About 13° S.E. of Aldebaran	Sl. B.T.	Nov. 2-3	E.
Leonids Brilliant shower	Nov. 13	Near (γ) Leonis In the Sickle	After midnight. v. Sw. Sk.	Nov. 12-14	N.E.
Leo Minorids	Nov. 16	Near (μ) Ursæ Maj., the Great Bear's hind feet	After 10 P.M. v. Sw. Sk.	Sept. 15, Oct. 16	N.
Andromedids. The Bielids. Fine display	Nov. 27	Near (γ) Andromedæ	Sl. T.	Nov. 17-23 Nov. 21-28	Overhead
Taurids	Nov. 30	Between Capella and (α) Persei	V. Sw.	Aug. 16 Sept. 15, Nov. 20	Overhead

Name of Shower	Date	Radiant Point	Characteristics	Other Dates Of Observation	Location
Zeta Taurids. Active shower in 1876	Dec. 6	Near the horns of the Bull	Sl. B.		E.
Geminids. Fine shower	Dec. 10	Near Castor	Sw.	Dec. 1-14	E.
Kappa Draconids	Dec. 22	Near Thuban (α) Draconis	Sw. Sk.	Nov. 14-23 Dec. 18-29	
Fire Ball Dates				Nov. 29 Dec. 2, 19, 21	

The Andromedes are usually red, sluggish in their movements, and leave only a small train.

Brilliant displays were seen in 1872 and 1885.

The Leonids are characterized by their exceedingly swift flight. They are of a greenish or bluish tint and leave behind them a vivid and persistent train. In most years the display is not especially noteworthy. Once in thirty-three years they afford an exhibition grand beyond description as in 1833 and 1866.

THE CONSTELLATIONS OF WINTER.

Map showing the principal stars visible from Lat. 40° N. at 9 o'clock, January first.

TAURUS (tâ´-rus)—THE BULL. (Face Southwest.)

LOCATION.—Taurus contains the well-known and unmistakable group the Pleiades, on the right shoulder of the Bull. A "V" shaped group known as the Hyades is just to the southeast of the Pleiades, in the face of the Bull, forming one of the most beautiful objects in the sky.

93

The brightest star in Taurus is Aldebaran, a ruddy-hued star known as "The Follower." It is at the beginning of the "V" in the Hyades, and is at the apex of a triangle formed by Capella, in Auriga, and α Persei, and equally distant from them both.

The star β called Nath, is peculiarly white, and is common to Taurus and Auriga. It represents the tip of one of the Bull's horns, and the right foot of the Charioteer. The Pleiades are mentioned in Chinese annals in 2357 B.C. On a photograph of the group over 2000 stars have been counted.

The ecliptic passes a little south of a point midway between the two horns, where a scattered and broken stream of minute stars can be seen.

Note two pretty pairs in the Hyades, one south of Aldebaran, the other northwest of it.

There are rich clusters below the tip of the horn over Orion's head.

Taurus was an important object of worship by the Druids.

Aldebaran is near one eye of the Bull, and used to be called "The Bull's Eye." An occultation of it by the moon, which not infrequently occurs, is a striking phenomenon.

The Eskimos regard the Pleiades as a team of dogs in pursuit of a bear. The group is receding from us at the rate of thirteen miles a second and has a common eastward motion of about ten seconds a century.

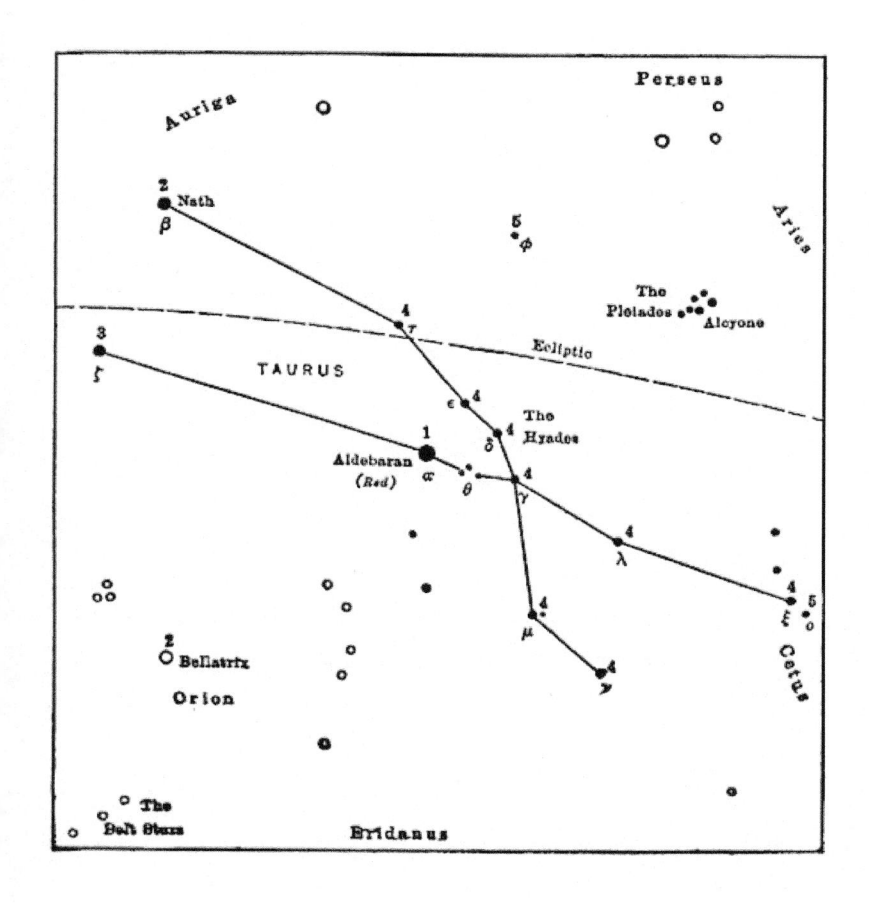

TAURUS

ORION (ŏ-rī'-on)—THE GIANT HUNTER. (Face South.)

LOCATION.—Orion is considered the finest constellation in the heavens. A line drawn from Nath to ζ Tauri (the tips of the Bull's horns), and extended 15°, strikes the brilliant Betelgeuze in Orion, known as the martial star. It forms the northeast corner of a conspicuous parallelogram. The splendid first-magnitude star Rigel is diagonally opposite Betelgeuze, and the girdle and sword of the Hunter lie within the parallelogram, a very striking group. The former is represented by three bright stars in a line 3° long known as the "Three Stars," because there are no other stars in the heavens that exactly resemble them in position and brightness.

In the sword there is the most remarkable nebula in the heavens. It may be seen with an opera-glass and in a telescope it is a wonderful sight. Bellatrix is called the Amazon star. Note the contrasting colours of α and β.

About 9° west of Bellatrix are eight stars in a curved line running north and south. These point out the Lion's skin held in the Hunter's left hand.

Below λ there are two stars forming a triangle with it. Flammarion calls this region the California of the sky.

The celestial equator passes nearly through δ.

Orion was worshipped in China during the one thousand years before our era, and was known to the Chinese as the "White Tiger."

The Eskimos see in the Belt stars the three steps cut by some celestial Eskimo in a steep snow bank to enable him to reach the top.

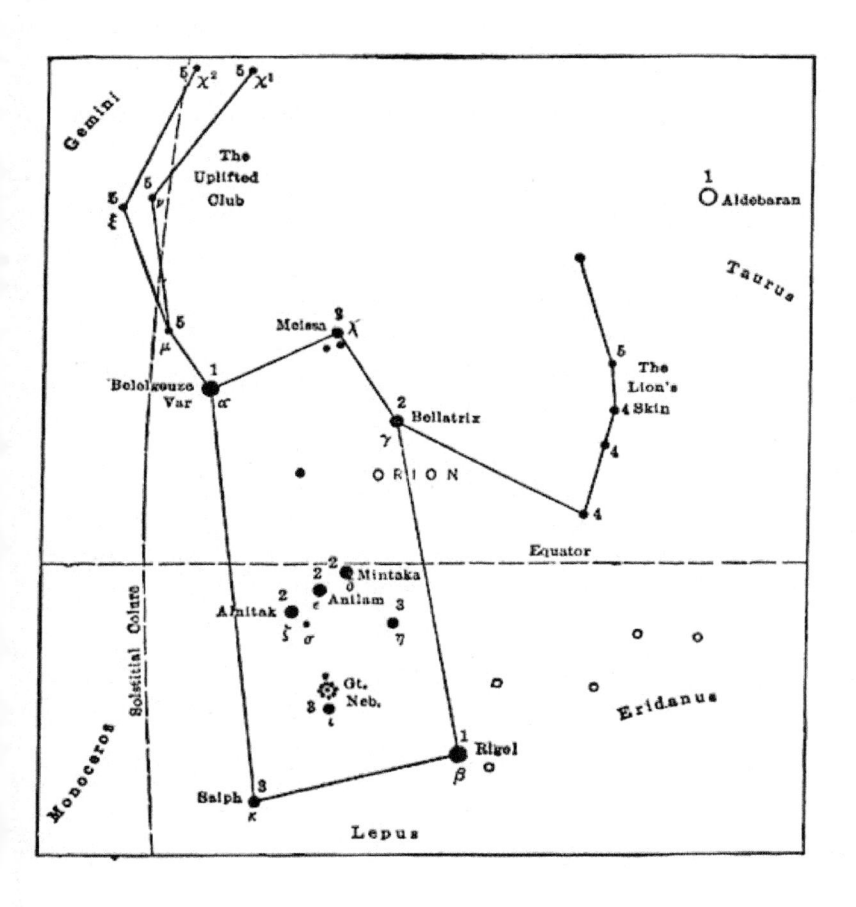

ORION

LEPUS (lē′-pus)—THE HARE. (Face South.)

LOCATION.—Lepus crouches under Orion's feet. Four stars in the constellation form an irregular and conspicuous quadrilateral.

γ is a beautiful double of a greenish hue.

Four or five degrees south of Rigel are four faint stars which are in the ear of the hare. They can be seen on a clear night with the naked eye.

The curved line of three stars θ, η, and ζ, are in the back of the hare.

Lepus is about 18° west of Canis Major, and, by reason of the earth's motion, the Great Dog seems to be pursuing the Hare around the heavens.

The first-magnitude stars that are visible in the winter season in this latitude present a fine contrast in color. Even the untrained eye can see a decided difference between the bluish white color of the brilliant Sirius, the Dog star that the Belt stars point south to, and Rigel, and the ruddy Betelgeuze. Procyon has a yellowish tinge and resembles the condition of our sun, while Betelgeuze is surrounded by heavy metallic vapors and is thought to be approaching extinction.

R marks the location of "Hind's crimson star," a famous variable.

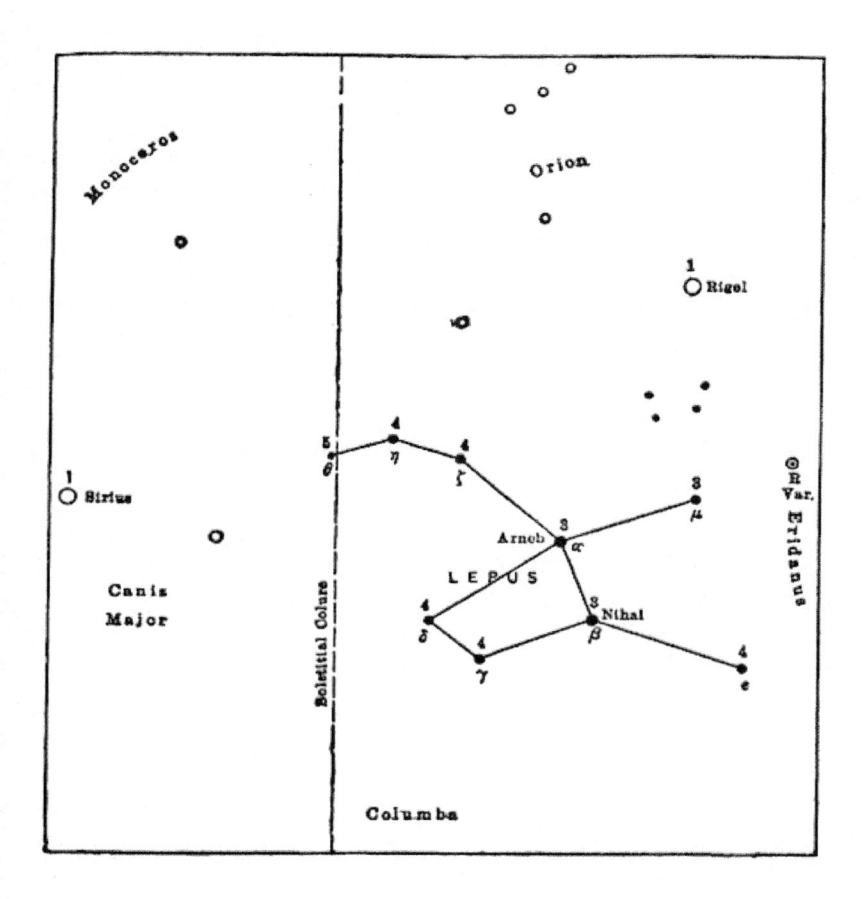

LEPUS

COLUMBA NOACHI (co-lum′-bä nō-ä′-ki)—NOAH'S DOVE. (Face South.)

LOCATION.—Columba is situated just south of Lepus. A line drawn from Rigel, in Orion, to β Leporis, and prolonged as far again, ends near α and β, the two brightest stars in Columba.

A line drawn from the easternmost star in the belt of Orion, 32° directly south, will point out Phaet, in Columba. It makes with Sirius, in Canis Major, and Naos, in the Ship, a large equilateral triangle.

The star β Columbæ may be known by means of a smaller star just east of it, marked γ.

The Chinese call α Chang Jin, the old Folks. Lockyer thinks it was of importance in Egyptian temple worship, and observed from Edfu and Philæ as far back as 6400 B.C.

On a clear starlight night there are not more than a thousand stars visible to the naked eye at one time. The largest telescope reveals nearly a hundred million.

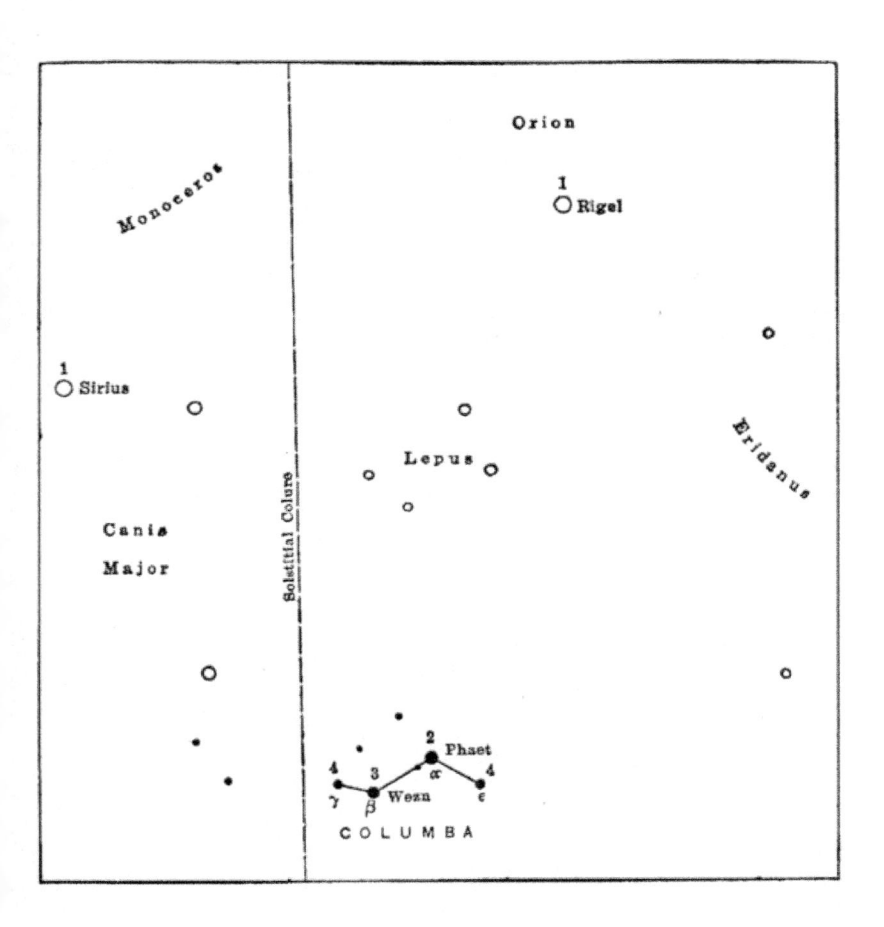

COLUMBA

CANIS MAJOR (kā'-nis mā-jor)—THE GREATER DOG.
(Face South.)

LOCATION.—The three stars in Orion's girdle point southeast to Sirius, the dog star, in Canis Major, the most brilliant star in the heavens. It was connected in the minds of the Egyptians with the rising of the Nile, and is receding from the earth at the rate of twenty miles a second.

The star ν is a triple. The cluster (41 M.) can be seen with an opera-glass, just below it.

Between δ and o^1 note a remarkable array of minute stars, also the very red star 22.

δ and ζ are doubles for an opera-glass.

Below η there is a fine group.

Betelgeuze, in Orion, Procyon, in Canis Minor, and Sirius form a nearly equilateral triangle. These stars with Naos, in the Ship, and Phaet, in the Dove, form a huge figure known as the Egyptian "X."

From earliest times Sirius has been known as the Dog of Orion. It is 324 times brighter than the average sixth-magnitude star, and is the nearest to the earth of all the stars in this latitude, its distance being 8.7 light years. At this distance the Sun would appear as a star a little brighter than the Pole Star.

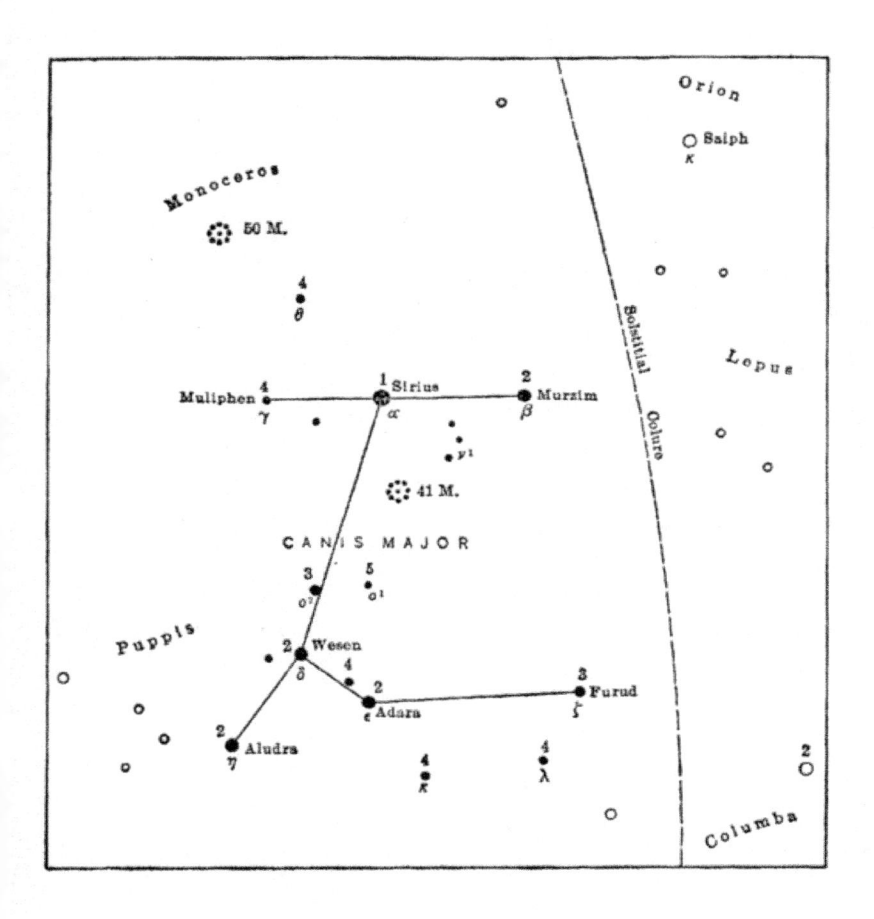

CANIS MAJOR

ARGO NAVIS (är'-go nā'-vis)—THE SHIP ARGO. (Face South.)

LOCATION.—Argo is situated southeast of Canis Major. If a line joining Betelgeuze and Sirius be prolonged 18° southeast, it will point out Naos, a star of the second magnitude in the rowlock of the Ship. This star is in the southeast corner of the Egyptian "X."

The star π is of a deep yellow or orange hue. It has three little stars above it, two of which form a pretty pair.

The star ζ has a companion, which is a test for an opera-glass.

The star κ is a double for an opera-glass.

Note the fine star cluster (46 M.).

The star Markeb forms a small triangle with two other stars near it.

The Egyptians believed that this was the ark that bore Osiris and Isis over the Deluge.

The constellation contains two noted objects invisible in this latitude, Canopus, the second brightest star, and the remarkable variable star η.

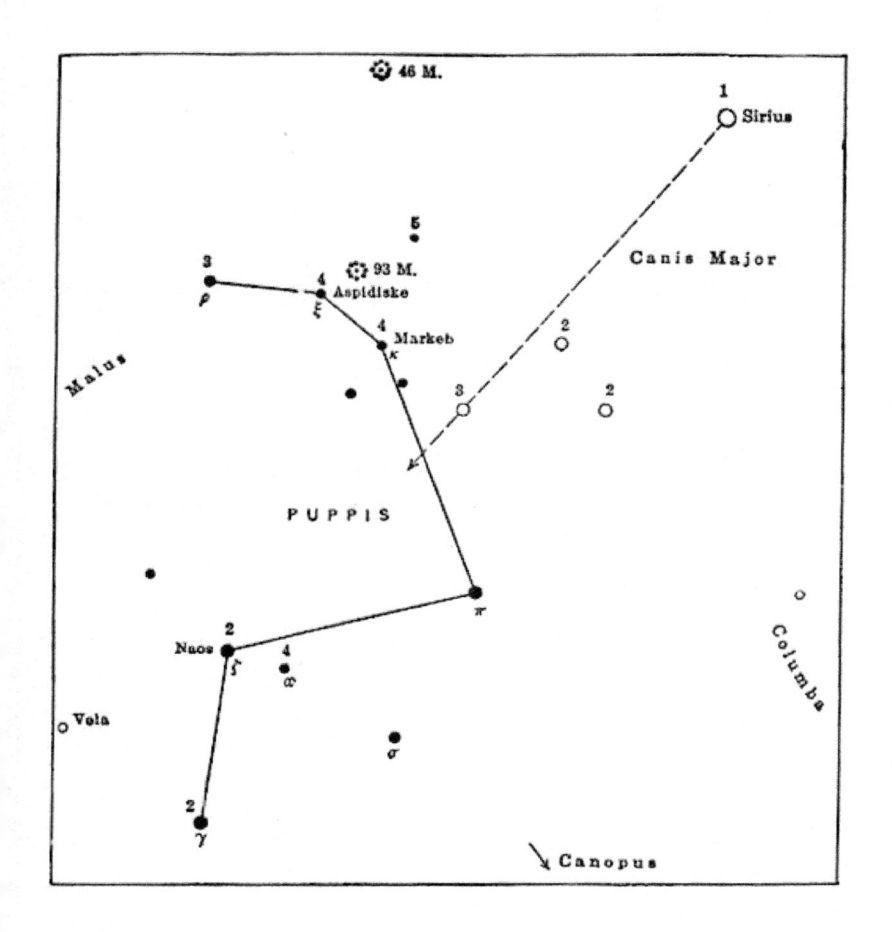

PUPPIS

MONOCEROS (mō-nos´-e-ros)—THE UNICORN. (Face South.)

LOCATION.—Monoceros is to be found east of Orion between Canis Major and Canis Minor. Three of its stars of the fourth magnitude form a straight line northeast and southwest, about 9° east of Betelgeuze, and about the same distance south of Alhena, in Gemini.

The region around the stars 8, 13, 17 is particularly rich when viewed with an opera-glass.

Note also a beautiful field about the variable S, and a cluster about midway between α and β.

Two stars about 7° apart in the tail of the Unicorn are pointer stars to Procyon. These stars are known as 30 and 31. The former is about 16° east of Procyon, and is easily identified as it has a sixth-magnitude star on either side of it. About 4° southwest of this star a good field-glass will reveal a beautiful star cluster.

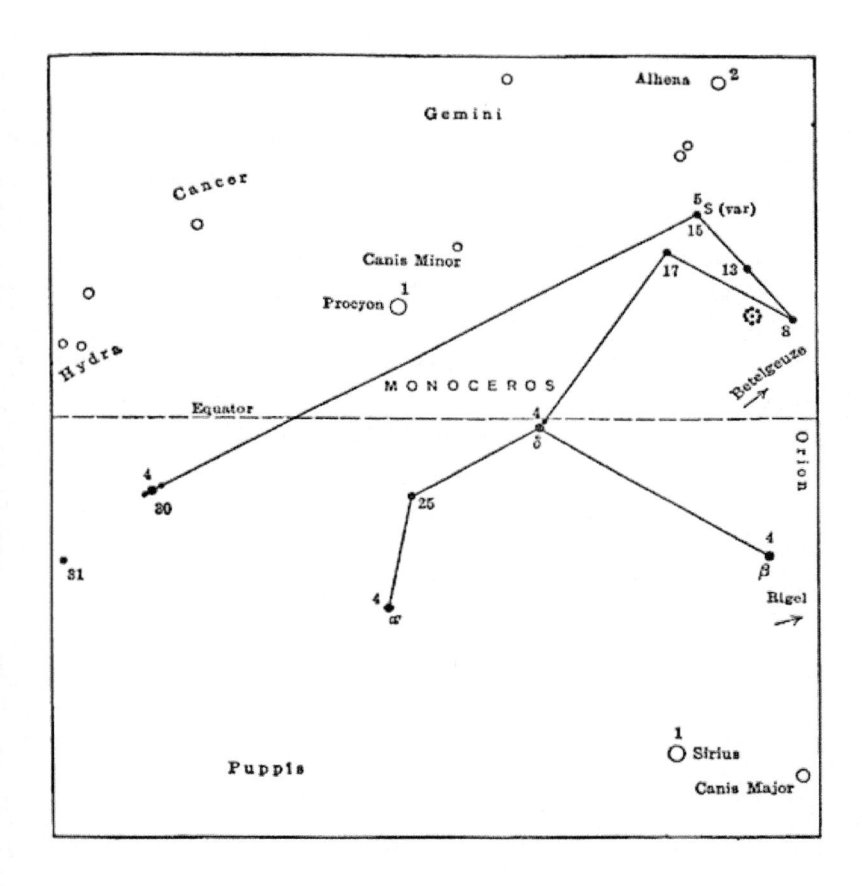

MONOCEROS

ERIDANUS (ē-rid′-a-nus)—OR THE RIVER PO. (Face Southwest.)

LOCATION.—Three degrees north and 2° west of Rigel, in Orion, lies β Eridani, the source of the River. Thence it flows west till it reaches π Ceti, then drops south 5°, thence east southeast, its total length being about 130°.

The great curve the River takes, just east of the Whale, resembles a horseshoe.

Acherna, the first-magnitude star in Eridanus, is too far south to be seen in this latitude.

Note the pretty star group around β and a pair of stars of an orange hue below ν.

The asterism known as "The Brandenburg Sceptre," consisting of four stars of the fourth and fifth magnitudes, can be seen arranged in a straight line north and south below the first bend in the River just west of Lepus.

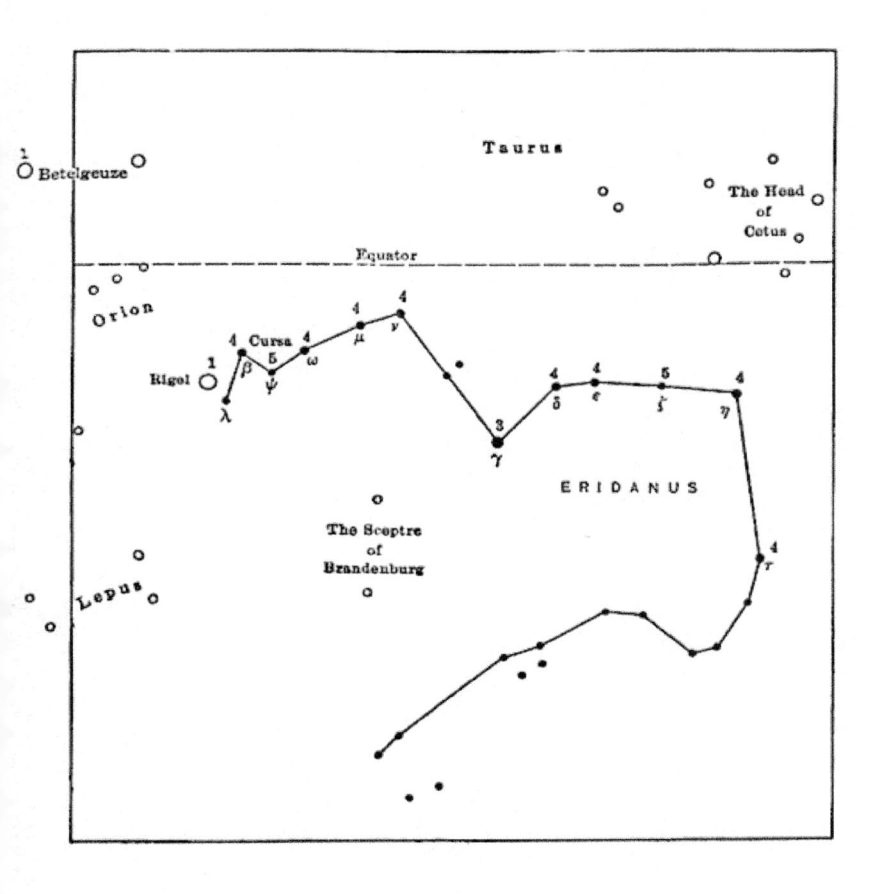

ERIDANUS

METEORIC SHOWERS.

JANUARY TO APRIL.

Name of Shower	Date	Radiant Point	Characteristics	Other Dates of Observation	Location
Quadrantids. Rich annual Shower	Jan. 2	(44) Boötis, between Boötes and Dragon's head	M. Sw. B.	Jan 3.	E.
Zeta Cancrids	Jan. 2-4	(ζ) Cancri, near Bee Hive			E.
Theta Ursids	Jan. 5	About 10° from β away from γ Ursæ Maj.	Small Sh. Sw. F.	Jan. 2-8	N.
Alpha Draconids	Feb. 1	Near Thuban α Draconis	Sl.	Jan. 9 Dec. 8	N.
in Alpha Aurigids	Feb. 7	Near Capella α Aurigæ	Sl.	Aug. 21 Sept. 12-22	High Southern Sky
Tau Leonids	Feb. 16	τ Leonis, Nov. 27 between Leo and Crater	Sl. Sk.	Dec. 12 Mar. 1-4	E.
Alpha Canum Ven. Well defined 1877	Feb. 20	Near Cor Caroli and Coma Berenices	V. Sw. B.		E.
A-β Perseids	Mar. 1	Between α and β Persei	V. Sl.	July—Dec. Mar. 13-19	N.W.
Beta Leonids or Beta Virginids	Mar. 14	Near Denebola β Leonis	Sl. B.	Mar. 3, 4 Dec. 12	S.E.

Other Dates

Name of Shower	Date	Radiant Point	Characteristics	of Observation	Location
Kappa Cepheids	Mar. 18	Near Polaris	Sl. B.	Oct. 4-17 Mar. 13-19	N.
Beta Ursids	Mar. 24	Near β Ursæ Maj.	Sw.	Apr. 10-16 Mar. 13-14 Dec. 2-9 Precise	N.
Zeta Draconids	Mar. 28	Near the Dragon's Head	Sl.	July 29 Aug. 24, etc.	N.

The Abbreviations under *Characteristics* are as follows:

V.—Very
Sh.—Short
M.—Moderately
B.—Bright
Sw.—Swift
F.—Faint
Sl.—Slow
Sk.—Streak leaving meteors
T.—Train leaving meteors

If you know the constellations, and memorize the following rhyme you will have ever at hand for reference at night, a reliable time-piece, a compass, and a perpetual calendar.

The numbers above the star names indicate consecutively the months of the year in which these respective objects rise about the first instant in the eastern sky. In addition to first-magnitude stars the rhyme refers to the head of Capricornus, the Sea Goat, the Great Square of Pegasus, and Orion's Belt. All except Arcturus rise between 9 and 9.30 P.M. Arcturus rises at 10 P.M., February 1st.

1
First Regulus gleams on the view,
2   3   4
Arcturus, Spica, Vega, blue,
5   6
Antares, and Altair,
7   8   9
The Goat's head, Square, and Fomalhaut,
10   11
Aldebaran, the Belt, a-glow,
12
Then Sirius most fair.

Eight months of the year are identified by the position of the Dipper at 9 P.M. In April and May it is north of the zenith. During July and August it is west of north. In October and November it lies close to the northern horizon and in January and February it is east of north with the pointers highest.

THE PLANETS.

It is not within the scope of this work to dwell at length on a discussion of the planets. Certain explanatory matter regarding them is necessary, however, to prevent confusion; for the student must bear in mind the fact that from time to time the planets appear in the constellations, and unless identified would lead him to think that the diagrams were inaccurate.

The reader is referred to any one of the four large plates that precede each season. He will observe that a portion of an ellipse has been traced on each of them, and that this line has been designated the Ecliptic, which simply means the sun's apparent pathway across the sky.

This pathway is divided into twelve equal parts of thirty degrees each, and to these twelve divisions are given the names of the constellations of the Zodiac in the following order: Aries (♈), Taurus (♉), Gemini (♊), Cancer (♋), Leo (♌), Virgo (♍), Libra (♎), Scorpio (♏), Sagittarius (♐), Capricornus (♑), Aquarius (♒), Pisces (♓).

The sun, starting from the first degree of Aries, the first day of spring, passes through one constellation a month. The planets follow the same pathway.

Confusion, therefore, respecting their identity can only arise in connection with a study of one of the twelve constellations named above, so that whenever a star of any size is seen in one of these constellations, not accounted for in the diagram, the student may conclude that this is a planet; especially if the unknown star does not twinkle. It now remains to identify the planet.

This can best be done by referring to an almanac, which states what planets are above the horizon, and which are morning and evening stars. By morning star is meant that the planet is east of the sun; by evening star, that it is west of the sun.

If the planet is in the west, and very brilliant, it is safe to assume that it is the planet Venus.

If it is brighter than any of the fixed stars, and is some distance from the sun, it is doubtless the colossal Jupiter.

If it is very red, it will probably be Mars.

Saturn is distinguished because of its pale, steady, yellow light.

As for Mercury, Uranus, and Neptune, the former is very near the sun, and seldom seen; while Uranus and Neptune are so inconspicuous as to lead to no confusion on the part of the novice.

A few notes of interest relative to the planets follow, taking them up in regular order passing outward from the sun: Mercury, Venus, Mars, Jupiter, Saturn, Uranus, Neptune.

MERCURY.

Mercury is the nearest to the sun of any of the planets. On this account, and because of its rapid changes, it is seldom seen.

The most favorable time for observing it is just after sunset, or just before sunrise, during the months of March, April, August, and September, when it may be seen for a few successive days.

The greatest distance it ever departs from the sun on either side varies approximately from sixteen to twenty-eight degrees. Its motion resembles a pendulum, swinging from one side of the sun to the other.

VENUS.

Venus approaches nearer to the earth and is more brilliant than any other planet. It is bright enough to cast a shadow at night, and is sometimes visible even at noonday. It is almost as large as the earth, and appears to oscillate, as Mercury does, on either side of the sun.

It never appears more than three hours after sunset, and as long before the sunrise, and is never more than forty-eight degrees from the sun.

MARS.

Mars is most like the earth of any of the planets, and, although not as interesting an object to view as the more brilliant planets, Venus and Jupiter, it claims our attention chiefly because of the surmises respecting its habitability.

Mars appears to the naked eye as a bright red star, and when at a favorable opposition to the earth (which occurs only once in every fifteen years) it rivals Jupiter in splendor.

The planet may be mistaken for the first magnitude stars, Antares in Scorpius, and Aldebaran in Taurus, near which it frequently passes.

The fixed stars, however, twinkle, while Mars glows steadily. If there is any doubt in the student's mind as to the identity of the planet, a few nights of observation, noting the changes in the planet's position, will decide the point. It takes Mars about fifty-seven days to pass through one constellation in the Zodiac.

JUPITER.

Jupiter is the largest of all the planets in the solar system, and it is easily distinguished from the fixed stars because of its brilliancy and splendor, exceeding in brightness all the planets excepting Venus, and casting a perceptible shadow.

It moves slowly and majestically across the sky, advancing through the Zodiac at the rate of one constellation yearly. It is therefore a simple matter to forecast its position, for, in whatever constellation it is seen to-day, one year hence it will be seen equally advanced in the next constellation.

Although Jupiter appears to move slowly, it really travels at the incomprehensible rate of five hundred miles a minute.

The most interesting feature about Jupiter for the amateur astronomer consists in observing four of its moons, which are visible with a small telescope. They appear like mere dots of light, and their transit of or occultation with the planet (that is, their disappearance before or behind its disk) can be watched, and is a never failing source of pleasure. A large telescope alone reveals Jupiter's four other moons.

SATURN.

Saturn is farther removed from the earth than any of the planets in the solar system, visible to the naked eye. It is distinguished from the fixed stars by the steadiness of its light, which is dull and of a yellow hue, though to some it appears to be of a greenish tinge. It seems barely to move, so slow is its motion among the stars, for it takes two and one half years to pass through a single constellation of the Zodiac.

Saturn has eight moons. Titan, its largest one, can be seen with a 3" glass. Its celebrated rings are telescopic objects but a small glass reveals them.

URANUS.

The student will hardly mistake Uranus for a fixed star, as it is only under the most favorable circumstances that it can be seen with the naked eye.

At its nearest approach to the earth, it is as bright as a sixth-magnitude star. Uranus is accompanied by four moons, and takes seven years to pass through a constellation of the Zodiac.

NEPTUNE.

Neptune is the most distant of the planets in the solar system, and is never visible to the naked eye.

The earth comes properly under a discussion of the planets, but a description of it is hardly within the scope of this work.

Confusion in identifying the planets is really confined to Mars and Saturn, for Venus and Jupiter are much brighter than any of the fixed stars, and their position in the heavens identifies them, as we have seen before.

The following table of first-magnitude stars in the Zodiacal constellations confines the question of identifying the planets to a comparison of the unknown star with the following-named stars:

Castor and Pollux in	Gemini.
Spica "	Virgo.
Regulus "	Leo.
Aldebaran "	Taurus.

Antares   "   Scorpius.

The first four stars named above are white in color, so that either Mars or Saturn is readily distinguished from them.

As for Aldebaran and Antares, which are both red stars, not unlike Mars and Saturn in color and magnitude, the fact that the latter do not twinkle, and that they do not appear in the diagrams, should satisfy the observer of their identity. Reference to an almanac, or a few nights of observation, will in any case set at rest any doubt in the matter.

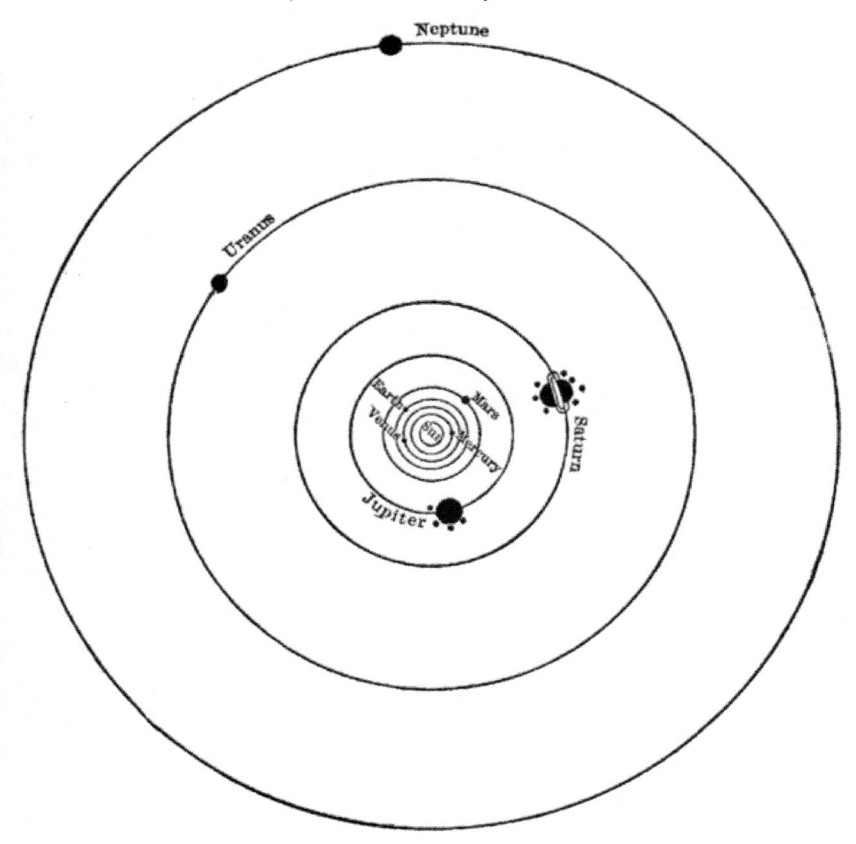

THE PLANETARY ORBITS

Jupiter

Saturn

Neptune

Uranus

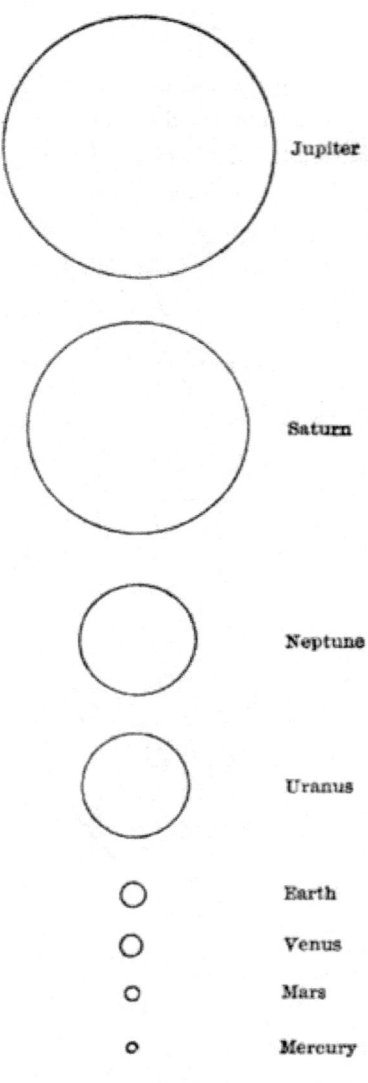

Earth
Venus
Mars
Mercury

COMPARATIVE SIZE OF THE PLANETS.

THE MILKY WAY.

The Milky Way, or Galaxy as it is sometimes called, is a great band of light that stretches across the heavens. Certain portions of it are worthy of being viewed with an opera-glass, which separates this seemingly confused and hazy stream into numberless points of light, emanating from myriads of suns.

This wonderful feature of the heavens is seen to best advantage during the months of July, August, September, and October. Beginning near the head of Cepheus, about thirty degrees from the North Pole, it passes through Cassiopeia, Perseus, Auriga, part of Orion, and the feet of Gemini, where it crosses the Ecliptic, and thence continues into the southern hemisphere, beyond our ken in these latitudes.

It reappears in two branches in the region of Ophiuchus, one running through the tail of Scorpius, the bow of Sagittarius, Aquila, Delphinus, and Cygnus; the other above and almost parallel to it, uniting with the first branch in Cygnus, and passing to Cepheus, the place of beginning.

The student should note especially the strange gap between α, γ, and ε Cygni. This dark space has been called the "Coal Sack."

The Milky Way in the vicinity of Cassiopeia is particularly rich, and well repays a search with an opera-glass.

"The Galaxy covers more than one tenth of the visible heavens, contains nine-tenths of the visible stars, and seems a vast zone-shaped nebula, nearly a great circle of the sphere, the poles being at Coma and Cetus."

THE MOTIONS OF THE STARS.

It may be that the student desires to proceed in this conquest of the sky at a more rapid pace than the scheme of study permits. To assist such, it should be borne in mind that the circumpolar constellations, as Ursa Major, Ursa Minor, Draco, Cepheus, and Cassiopeia, are designated,—are visible in our latitude in the northern sky every night.

A reference to their diagrams, and a glance at any of the large plates showing the entire group in their respective positions, will suffice for the student to identify them.

The hours of darkness alone limit the speed with which a knowledge of the constellations can be acquired.

Let us suppose that the student begins his search for the constellations on the night of April 1st, at nine P.M. He has for his guide the large plate, and the spring group of eleven constellations set forth in the diagrams. The remaining three constellations of the circumpolar group are, as we have seen before, visible in the north.

If he faces the western sky, he will see Andromeda just setting, and Perseus, Taurus, Orion, Lepus, and Canis Major but a short distance above the horizon. If he is so fortunate as to be able to identify these, and the spring group, he may turn his attention wholly to the eastern sky, where new constellations await him.

In the southeast he may see Virgo. In the east well up blazes Arcturus, the gem of Boötes, below which is the beautiful Northern Crown, with the diamond in the head of Serpens beneath it. Hercules is rising, and Vega in the Lyre should be seen just flashing on the view in the northeast.

This completes the list of wonders visible at this precise time, but the stars apparently are never still, and doubtless, while the student has been passing from one constellation to another in the western and southern skies, others have been rising in the east and northeast.

At ten P.M. the Lyre is well up, and Ophiuchus and Libra can be discerned. At midnight Scorpius and Cygnus are ready to claim the attention. By two o'clock A.M., Aquila, Delphinus, and Sagittarius have risen, and at break of day Andromeda, Pegasus, and Capricornus can be seen if the student has had the courage to remain awake this length of time.

In no way can the seeming movement of the stars be better understood than by actual observation. The observer must bear in mind that the movement is an apparent one: that it is the earth that is moving and not the stars. He has only to think of the analogy of the moving train beside the one that is standing still, and the true state of affairs will at once be evident.

To further appreciate this apparent change in the situation of the constellations, the student should refer to the large plates successively. In each successive one he will note the advancement westward of the constellations mentioned above, rising in the east late at night.

The student can best get an idea of this westward apparent movement of the stars by noting the position of some bright first-magnitude star from night to night. He will soon be able to calculate the position of this star a month or more ahead, and this calculation applies to all the constellations and stars.

It is not within the scope of this work to go into this matter in detail. The author merely desires to mention this fact of apparent change of position in the stars, a fact that will be noticeable to the observer in a short time, and a fact that it is hoped he will be able to explain to his own satisfaction with the aid of the foregoing remarks.

It will be noticed that the stars on the diagrams are all numbered and lettered. The numbers refer to the magnitude of the star,—that is, the brightness of it, the first-magnitude stars being the brightest, the second-magnitude stars two-and-a-half times less bright, etc.

The letters are those of the Greek alphabet, and the student if not familiar with it is advised to consult a Greek grammar.

In the text, in referring to certain stars in the constellations, the genitive case of the Latin name of the constellation is given; for example, Vega is known as α Lyrae, meaning alpha of Lyra, Aldebaran as α Tauri, alpha of Taurus, etc.

The twilight hour affords an excellent opportunity of fixing the relative positions of the first-magnitude stars in the mind, for at that time they alone, save the planets, are visible.

METEORS, OR SHOOTING-STARS.

As this work is designed primarily to cover what is observable in the starlit heavens with the naked eye, the subject of meteors, or shooting-stars, comes properly within its scope.

There are few persons, if any, who have not witnessed the sight of a splendid meteor speeding across the sky, and such a sight always calls forth exclamations of wonder and delight.

Apparently these evanescent wanderers in space are without distinctive features, and baffle classification; but, like all that nature reveals to us, they have been found, for the most part, to conform to certain laws, and to bear certain marks of resemblance that permit of their identification and classification.

By careful observation for over fifty years the meteors, generally speaking, have been so arranged that they come under the head of one of the nearly three hundred distinct showers which are now recognized by astronomers.

Many of these showers are too feeble and faint to be worthy of the attention of one not especially interested in the subject, but certain ones are well worth observing. There is always a pleasure in being able to recognize at a glance a certain definite manifestation of nature, be it a rare flower or a flashing meteor.

The generally accepted theory respecting the meteors is that they were all originally parts of comets now disintegrated, and the four well-known showers of April 20th, August 10th and 14th, and November 27th, bear testimony to this theory.

The apparent velocity of the meteors is between ten and forty-five miles a second, and their average height is about seventy-six miles at first appearance, and fifty-one miles at disappearance. Occasionally a meteor is so large and compact as to escape total destruction, and falls to the earth. Specimens of these meteorites are to be found in our best museums.

I have seen fit to divide the principal meteor showers into four groups, according to the seasons in which they appear, and have placed them respectively at the conclusion of each season's work on the constellations.

By radiant point is meant the point from which the meteors start on their flight. This point is an apparent one, however, due to an illusion of perspective, for the meteors really approach the earth in parallel paths.

The dates given for these showers are those of the maxima, and the meteors should be looked for several nights before and after the dates specified.

The showers that are to be seen after midnight are, unless of special note, omitted.

There are, besides the meteors that have been classified, certain shooting-stars that apparently have no determined radiant point. These are called sporadic meteors.

In these lists of meteors, the radiant point is only approximately given; for scientific purposes a far more exact position is required in terms of right ascension and declination. There are several good lists of meteoric showers to be obtained, which afford this information for those who care to pursue the matter more in detail. See the Rev. T.W. Webb's book, entitled *Celestial Objects for Common Telescopes*. For purposes of identification, the radiant points here given will be found for the most part sufficient.

NOTE.

Many readers of this book may be the fortunate possessors of small telescopes. It may be that they have observed the heavens from time to time in a desultory way and have no notion that valuable and practical scientific research work can be accomplished with a small glass. If those who are willing to aid in the great work of astrophysical research will communicate with the author he will be pleased to outline for them a most practical and fascinating line of observational work that will enable them to share in the advance of our knowledge respecting the stars. It is work that involves no mathematics, and its details are easily mastered.

THE NAMES OF THE STARS AND THEIR MEANINGS.

ACUBENS, α *Cancri*, "the claws."

Situated in one of the Crab's claws. It is white in color and culminates[1] March 18th.

A-DAR'-A, ε *Canis Majoris*, "the virgins," a name for four stars, of which Adara is brightest.

Situated in the Dog's right thigh. It is pale orange in color, and culminates Feb. 11th.

ADHIL, ε *Andromedæ*, "the train of a garment."

Situated in the left shoulder of the chained lady.

ALADFAR (al-ad-fär), μ *Lyræ*, "the talons" (of the falling eagle)

AL BALI, ε *Aquarii*, "the good fortune of the swallower."

AL-BI'-REO, or AL-BIR'Ë-O β *Cygni*, origin doubtful. Means the beak of the hen.

Situated in the beak of the Swan and the base of the Cross.

Its color is topaz yellow, and it culminates Aug. 28th.

ALCAID, η *Ursæ Majoris*. *See* Benetnasch.

ALCHIBA (al-kē-bä'), α *Corvi*, "the tent," the desert title for the constellation.

Situated in the eye of the Crow. Orange in color.

ALCOR (al'-kôr), g *Ursæ Majoris*, "the cavalier" or "the rider."

Situated close to Mizar in the handle of the "Dipper." Silver white in color. The Arabs called this star "Saidak," meaning "the proof," because they used it to test a good eye.

AL-CY'-O-NE, η *Tauri*.

---

[1] It will be noted that the date of culmination is given in almost every case. By culmination is meant the highest point reached by a heavenly body in its path, at which point it is said to be on the meridian. In this hemisphere this is in each case the highest point North.

For example:—the culmination of the sun occurs at noon.

The time when the stars here mentioned culminate on the dates specified is in each case nine o'clock P.M.

Greenish yellow in color. The brightest of the Pleiades.

Situated in the neck of the Bull.

AL-DEB'-A-RAN, α *Tauri*, "the hindmost" or the "follower," *i.e.* of the Pleiades.

Situated in the eye of the Bull. Pale rose in color. It is receding from the earth at the rate of thirty miles per second, and culminates Jan. 10.

α Tauri is sometimes called Palilicium.

ALDERAMIN (Al-der-am'-in), α *Cephei* "the right arm." It now marks the shoulder of Cepheus.

White in color. It culminates Sept. 27th.

ALDHAFERA, ζ *Leonis*.

Situated in the "Sickle," and the neck of the Lion. It culminates April 8th.

ALFIRK (al-ferk'), or ALPHIRK, "stars of the flock," β *Cephei*.

The Arab name for the constellation. Situated in the girdle of Cepheus. White in color. It culminates Oct. 2d.

ALGEIBA (al-jē'-bä), γ *Leonis*, "the mane."

Situated in the "Sickle," and the shoulder of the Lion. It is approaching the earth at the rate of twenty-four miles per second, and culminates April 9th.

AL'-GE-NIB, γ *Pegasi*, "the wing," possibly the "flank" or "side."

Situated in the wing of the Horse. White in color, and culminates Nov. 14th.

AL'-GE-NIB, α *Persei*, "the side," or Mirfak, "the elbow."

Situated in the right side of Perseus. Lilac in color and approaching the earth at the rate of six miles per second. It culminates Jan. 1st. This star is also called Alchemb.

ALGENUBI (al-je-nö'-bi), ε *Leonis*, "the head of the Lion."

A yellow star situated in the Lion's mouth.

AL'-GOL, β *Persei*, "the ghoul" or "demon."

Situated in the head of the Medusa held in the Hero's left hand. White in color. It is approaching the earth at the rate of one mile per second, and culminates Dec. 23d.

ALGORAB (al-go-räb'), or ALGORES, (δ) *Corvi*, "the raven."

Situated on the right wing of the Crow. Pale yellow in color. It culminates May 14th.

ALHENA (al-hen'-a), γ *Geminorum*, "a brand on the right side of the camel's neck," or a "ring" or "circlet."

Situated in the left foot of Pollux. White in color, and culminates Feb. 8th. Alhena is sometimes called Almeisam.

AL-I-OTH, ε *Ursæ Majoris*, disputed derivation.

Situated in the tail of the Great Bear. It is approaching the earth at the rate of nineteen miles per second. It culminates May 20th. Alioth, the name sometimes given to α and θ Serpentis.

AL-KAID (al-kād), *See* ALCAID.

ALKALUROPS (al-ka-lū'-rops), μ *Boötis*, "a herdsman's club, crook, or staff."

Situated near the right shoulder of the Herdsman. Its color is flushed white.

ALKES (al'-kes), α *Crateris*, from Al Kas, "the cup," the Arab name for the constellation.

Situated in the base of the Cup. Orange in color, and culminates April 20th.

ALMAC, γ *Andromedæ*, "a badger," possibly "the boot."

Situated in the left foot of Andromeda. Orange in color, and culminates Dec. 8th.

AL NAAIM, τ and ν Pegasi, "the cross bars over a well."

AL NASL (al-nas'l), or ELNASL (el-nas'-l), γ *Sagittarii*, "the point head of the arrow."

Situated in the arrow's tip. It is yellow in color, and culminates Aug. 4th. This star sometimes called Nushaba and Warida.

AL NATH, or NATH γ *Aurigæ*, and β *Tauri*, "the heel of the rein-holder," the "butter" *i.e.* the "horn."

Situated in the right foot of the Charioteer, and the tip of the northern horn of the Bull. Brilliant white in color, and culminates Dec. 11th.

ALNILAM (al-ni-lam'), ε *Orionis*, "a belt of spheres or pearls."

Situated in Orion's belt. It is bright white in color, and is receding from the earth at the rate of sixteen miles per second. It culminates Jan. 25th.

ALNITAK (al-ni-tak'), ζ *Orionis*, "the girdle."

Situated in Orion's belt. Topaz yellow in color. It is receding from the earth at the rate of nine miles per second, and culminates Jan. 26th.

AL-NIYAT, σ *Scorpii*, "the outworks of the heart."

Situated near the Scorpion's heart. It is creamy white in color.

AL'-PHARD, or (al-färd'), α *Hydræ*, "the solitary one in the serpent."

Situated in the heart of Hydra. Orange in color, and culminates Mar. 26th. The Chinese called this star "the Red Bird."

AL-PHEC'-CA, α *Coronæ Borealis*, "the bright one of the dish." *See* Gemma. Century Dictionary gives meaning "the cup or platter of a dervish."

AL'-PHE-RATZ or (al-fe-rats'), α *Andromedæ*, "the head of the woman in chains." "The navel of the horse."

Situated in the head of Andromeda. White and purplish in color. It culminates Nov. 10th. Alpheratz is some times called Sirrah.

AL-PHIRK, β *Cephei*, from al-Firk, the flock.

AL RAKIS, μ *Draconis*, "the dancer."

Situated in the Dragon's nose. Brilliant white in color. The Century Dictionary gives for this star Arrakis, "The trotting camel."

AL RESCHA, α *Piscium*, "the cord or knot."

Situated in the knot joining the ribbons that hold the Fishes together. Pale green in color, and culminates Dec. 7th.

ALSAFI σ *Draconis*.

ALSCHAIN (al-shān'), β *Aquilæ*, part of the Arab name for the constellation.

Situated in the head of Antinoüs. Pale orange in color, and culminates Sept. 3d.

AL SHAT, ν *Capricorni*, "the sheep."

AL'-TAIR, or ATAIR, α *Aquilæ*, "the flying eagle," part of the Arab name for the constellation.

Situated in the neck of the Eagle. Yellow in color, and culminates Sept. 1st.

ALTERF (al-terf'), λ *Leonis*, "the glance," *i.e.* the Lion's eye.

Situated in the Lion's mouth, the point of the Sickle. Red in color.

ALUDRA (al-ö'-dra), η *Canis Majoris*, "the virgins." The four stars near each other in Canis Major.

Situated in the Great Dog's tail. Pale red in color, and culminates Feb. 21st.

ALULA BOREALIS, ν *Ursæ Majoris*.

ALULA AUSTRALIS, ζ *Ursæ Majoris* The "northern and southern wing."

Situated in the Southern hind foot of the Great Bear. The latter star is sometimes called El Acola.

AL'-YA, θ *Serpentis*.

Situated in the tip of the Serpent's tail. Pale yellow in color. It culminates Aug. 18th.

ANCHA, θ *Aquarii*, "the hip."

Situated in the right hip of Aquarius.

ANT-ĀR-ES, or AN-TA'-REZ, α *Scorpii*, "the rival of Mars."

Situated in the heart of the Scorpion. Fiery red and emerald green in color. It culminates July 11th.

ARC-TŪ-RUS, α *Boötis*, "the leg of the lance-bearer," or "the bear-keeper."

Situated in the left knee of the Herdsman. Golden yellow in color. It culminates June 8th.

ARIDED, *See* DENEB.

ARNEB (är'-neb), α *Leporis*, "the hare," the Arab name for the constellation.

Situated in the heart of the Hare. Pale yellow in color. It culminates Jan. 24th. α *Leporis* is sometimes called Arsh.

ARKAB (är'-kab), β *Sagittarii*, "the tendon uniting the calf of the leg to the heel."

Situated in the Archer's left fore leg.

ASHFAR, μ and ε *Leonis*, "the eyebrows."

Situated close to the Lion's right eye. μ orange in color, sometimes called Alshemali or Asmidiske.

ASPIDISKI (as-pi-dis′ke), or ASMIDISKE, ι *Argus*, "in the gunwale."

Situated in the shield which ornaments the vessel's stern. Pale yellow in color. The Century Dictionary gives "a little shield" as the meaning for this star name.

ASCELLA, ζ *Sagittarii*, "the armpit."

Situated near the Archer's left armpit. It culminates Aug. 19th.

ASCELLUS, θ *Boötis*.

It marks the finger tips of the Herdsman's upraised hand.

ASCELLUS BOREALIS, ν *Cancri*, "northern ass."

Straw color.

ASCELLUS AUSTRALIS, δ *Cancri*, "the southern ass."

Situated on the back of the Crab. Straw color.

ATIK, o *Persei*.

Situated in the wing on the right foot of Perseus.

AZELFAFAGE, π *Cygni*, "the horse's foot or track."

AZHA, η *Eridani*, "the ostrich's nest."

Pale yellow in color.

BAHAM, θ *Pegasi*, "the young of domestic animals."

Situated near the left eye of Pegasus.

BAT′EN KAITOS, ζ *Ceti*, "the whale's belly."

A topaz-yellow-colored star, which culminates Dec. 5th.

BEID (bā′-id), o *Eridani*, "the egg."

A very white star.

BEL′-LA-TRIX, γ *Orionis*, "the female warrior." The Amazon star.

Situated in the left shoulder of Orion. Pale yellow in color. It is receding from the earth at the rate of six miles per second, and culminates Jan. 22d. The Century Dictionary gives the color as very white.

BE-NET′-NASCH, η *Ursæ Majoris*, "the chief or governor of the mourners" (alluding to the fancied bier).

Situated in the tip of the Great Bear's tail. Brilliant white in color. It is approaching the earth at the rate of sixteen miles per second, and culminates June 2d. This star is also called Alkaid, from al-kaid, "the Governor."

BETELGEUZE (BET-EL-GERZ'), α *Orionis*, "the giant's shoulder," or "the armpit of the central one."

Situated in the right shoulder of Orion. Orange in color. It is receding from the earth at the rate of ten miles per second, and culminates Jan. 29th.

Sometimes called Mirzam, the roarer.

BOTEIN (bō-tē-in'), δ *Arietis*, "the little belly."

CA-PEL'-LA, α *Aurigæ*, "the she-goat."

Situated in the left shoulder of the Charioteer. It is a white star, and is receding from the earth at the rate of fifteen miles per second. It culminates Jan. 19th. The color of Capella is nearly that of the sun.

CAPH (kaf), β *Cassiopeiæ*, "the camel's hump," or "the hand."

It is white in color, and culminates Nov. 11th.

CAS'-TOR, α *Geminorum*, "the horseman of the twins."

Its color is bright white, and it culminates Feb. 23d. Situated in the head of Castor. The Century Dictionary gives the color as greenish.

CHELEB, β *Ophiuchi*, also CEB'ELRÁI from kelb, the shepherd's dog.

Situated in the head of the Serpent. It is a yellow star, and culminates Aug. 30th.

CHORT (chôrt), θ *Leonis*.

Situated in the hind quarters of the Lion. It culminates April 24th. The Century Dictionary has θ *Centauri* for this star.

COR CAROLI (kôr kar'-ō-lī), α *Can. Ven.*, "the heart of Charles II."

It is flushed white in color, and culminates May 20th. A yellowish star according to the Century Dictionary.

CUJAM, ω *Herculis*. Word used by Horace for the club of Hercules.

CURSA (KER' SA), β *Eridani*, "the footstool of the central one," or "the chair or throne."

Situated about at the source of the river near Orion. Topaz yellow in color, and culminates January 13th. This star is also known as Dhalim (Tha′lim) ("the ostrich").

DABIH (dä′-be), β *Capricorni*, "the lucky one of the slaughterers," or "the slayer's lucky star."

Situated in the head of the Sea-Goat. It is an
orange-colored star, and culminates Sept. 10th.

DĒ′-NEB, or ARIDED (ar′-i-ded), α *Cygni*, "the hen's tail," "the hindmost."

Situated in the tail of the Swan, and at the top of the Cross. Brilliant white in color. It is approaching the earth at the rate of thirty-six miles per second. It culminates Sept. 16th.

DENEB AL OKAB (den′-eb al-ō-kâb), ε and ζ *Aquilæ*, "the eagle's tail."

DENEB ALGEDI (den′-eb al′-jē-dē), δ *Capricorni*, "the tail of the goat."

Situated in the tail of the Sea-Goat.

DENEB AL SHEMALI (den′-eb-al-she-mä-le), ι *Ceti*

A bright yellow star situated at the tip of the northern fluke of the monster's tail.

DENEB KAITOS (den′-eb kī′-tos), β *Ceti*, "the tail of the whale."

Situated in the tail of the Whale. It is a yellow star, and culminates Nov. 21st. This star sometimes called Diphda.

DE-NEB′-O-LA, β *Leonis*, "the lion's tail."

It is a blue star which is approaching the earth at the rate of twelve miles per second. It culminates May 3d. This star also called Dafirah, and Serpha.

DSCHUBBA, δ *Scorpii*, "the front of the forehead."

Situated in the head of the Scorpion. It culminates July 4th.

DSIBAN, ψ *Draconis*.

Pearly white in color.

DUB′-HE (döb′-he), α *Ursæ Majoris*, "a bear."

The northern pointer star. It is a yellow star, and is approaching the earth at the rate of twelve miles per second. It culminates April 21st. The Arabs called the four stars in the Dipper the "bier."

EL NATH β *Tauri*, the one who butts. This star is receding at the rate of five miles a second.

ELTANIN, or ETANIN (et'-ā-nin), γ *Draconis*, "the dragon," "the dragon's head."

It is orange in color and culminates Aug. 4th. Rasaben is another name for this star.

E'-NIF, or en'-if, ε *Pegasi*, "the nose."

Situated in the nose of Pegasus. It is a yellow star, which is receding from the earth at the rate of five miles per second, and culminates Oct. 4th. This star was also called fum-al-far-as, "the mouth of the horse."

ER RAI (er-rā'-ē), γ *Cephei*, "the shepherd."

Situated in the left knee of Cepheus. It is yellow in color and culminates Nov. 10th.

FOMALHAUT (Fō'-mal-ō), (disputed pronunciation), α *Piscis Austri*, "the fish's mouth."

Situated in the head of the Southern Fish. It is reddish in color, and culminates Oct. 25th. This star was also known as the first frog, the second frog being β Ceti.

FURUD, or PHURUD (fu-rōd), ζ *Canis Majoris*, "the bright single one."

Situated in the left hind paw of the Greater Dog. It is light orange in color.

GEM'-MA, α *Coronæ Borealis*, "a bud."

The brightest star in the Northern Crown. It is brilliant white in color, and is receding from the earth at the rate of twenty miles per second. It culminates June 28th. This star is also known as Alphecca and Alfeta.

GIANSAR λ *Draconis*, "the twins," "the poison place."

Situated in the tip of the Dragon's tail. An orange-colored star. It culminates April 28th.

GIEDI, α *Capricorni*, also called Algied'-i, the goat.

Situated in the head of the Sea-Goat. It is a yellow star, and culminates Sept. 9th.

GIENAH, γ *Corvi*, "the right wing of the raven."

Situated in the Crow's wing. It culminates May 10th.

GIENAH, ε *Cygni*, "the wing."

Situated in the Swan's wing. It is a yellow star, and culminates Sept. 17th.

GOMEISA (gō-mī'-zä), GOMELZA, β *Canis Minoris*, "Watery-eyed, weeping." A white star.

Situated in the neck of the Lesser Dog.

GRAFFIAS, β *Scorpii*, derivation unknown; the name may mean "the crab." This star was also called Ak'rab, the Scorpion.

Situated in the head of the Scorpion. It is a pale white star, and culminates July 5th.

GRUMIUM (grö'-mi-um), ξ *Draconis*, "the dragon's under jaw." A yellow star.

HAM'-AL or (ha-mäl'), α *Arietis*, "the head of the sheep."

Situated in the forehead of the Ram. It is yellow in color, and is approaching the earth at the rate of nine miles per second. It culminates Dec. 11th.

HOMAM (ho-mam'), ζ *Pegasi*, "the lucky star of the hero, or the whisperer."

Situated in the neck of Pegasus. Light yellow in color. It culminates Oct. 22d. The Century Dictionary gives this star name to η *Pegasi*.

HYADUM I, γ *Tauri*.

Situated in the Hyades, the nose of the Bull. A yellow star.

IZAR (ē-zär), Mirach, or Mizar, ε *Boötis*, "the girdle."

Pale orange in color. It is approaching the earth at the rate of ten miles per second, and culminates June 16th. A beautiful colored double star.

JABBAH (Jab'-bä), ν *Scorpii*, "crown of the forehead."

A triple star.

KAUS (kâs), AUSTRALIS, ε *Sagittarii*, "the southern part of the bow."

An orange-colored star. It culminates Aug. 8th.

KAUS (kâs), BOREALIS, λ *Sagittarii*, "the northern part of the bow."

Orange color.

KITALPHA, α *Equulei*, the Arab name for the asterism. In the head of the Little Horse. It culminates Sept. 24th.

KO'-CHAB (kō-käb'), β *Ursæ Minoris*, "the star of the North." Situated in the right shoulder of the Little Bear. One of the two Guardians of the Pole. It is reddish in color, and is receding from the earth at the rate of eight miles per second. It culminates June 19th.

KORNEPHOROS, β *Herculis*, the Arab name for the constellation.
Situated in the right arm-pit of Hercules. Pale yellow in color. It is approaching the earth at the rate of twenty-two miles per second. It culminates July 12th.

LESUTH, ν *Scorpii*, "the sting."
Situated in the tip of the Scorpion's tail. It culminates July 27th.

MARFAK (mär'fak), θ *Cassiopeiæ*, "the elbow."
Situated in the left elbow of Cassiopeia. This star name is also given to μ Cassiopeiæ.

MARFIC (mär'-fik), λ *Ophiuchi*, "the elbow."
Situated in the left elbow of the Serpent Bearer. Yellowish white in color.

MARFIK (mär' fik), or MARSIC, κ *Herculis*, "the elbow."
Situated in the right elbow of Hercules. Light yellow in color.

MAR'-KAB (mär'kab), α *Pegasi*, Arab word for "saddle". Century Dictionary gives "a wagon" or "chariot."
Situated in the base of the Horse's neck. It is a white star which is receding from the earth at the rate of three quarters of a mile a second. It culminates Nov. 3d.

MARKEB, κ *Argus*.
Situated in the stern of the Ship. It culminates Mar. 25th.

MARSYM, λ *Herculis*, "the wrist."
Situated in the left wrist of Hercules. Deep yellow in color.

MATAR or SAD (Säd), "a lucky star," or more fully, Sad-Mator, η *Pegasi*, "the fortunate rain."
Situated in the left fore leg of Pegasus.

MEBUSTA, MEBSUTA (Meb-sö'-ta), or MEBOULA, ε *Geminorum*, "the outstretched."

A brilliant white star situated in the right knee of Castor.

MEDIA, or KAUS MEDIA, δ *Sagittarii*, "middle (of the) bow." Orange yellow in color. It culminates Aug. 8th.

MĒ-GRES, or (Mē'-grez), δ *Ursæ Majoris*, "the root of the bear's tail."

It is a pale yellow star, and culminates May 10th. This star is the faintest of the seven which form the Dipper.

MEISSA, λ *Orionis*.

Situated in the face of the Giant Hunter. Pale white in color.

MEKBUDA (mek-bū'-dā), ζ *Geminorum*, "the contracted (arm)." Situated in the left knee of Pollux. Pale topaz in color.

MENKALINAN (men-ka-lē-nan' or Men-kal'-i-nan), β *Aurigæ*, "the shoulder of the rein-holder or driver."

Situated in the right arm of the Charioteer. A lucid yellow star which is receding from the earth at the rate of seventeen miles per second. It culminates Jan. 29th. This star was one of the first discovered and most remarkable "spectroscopic binaries."

MENKAR (men'kär), α *Oeti*, "the nose, or snout."

Situated in the nose of Cetus. Bright orange in color. It culminates Dec. 21st. Sometimes written Menkab.

MENKIB, ξ *Persei*, "the shoulder."

Situated in the calf of the right leg of Perseus.

MERAK (mē'rak), β *Ursæ Majoris*, "the loin of the bear."

A greenish white star which is approaching the earth at the rate of eighteen miles per second. It culminates Apr. 20th. The southern of the two "pointers."

MESARTIM (mē-sär'tim), γ *Arietis*, the Hebrew word for "minister."

Situated in the Ram's left horn. Bright white in color.

MINTAKA (min'ta-kä), δ *Orionis*, "the belt (of the giant)."

One of the three stars in Orion's belt. A brilliant white star with very little motion. It culminates Jan. 24th.

MĪ'-RA (mī'ra or mē'ra), o *Ceti*.

Situated in the neck of Cetus. A famous variable, flushed yellow in color. It culminates Dec. 15th.

MĪ'RACH, or MIRAK (mī'rak or mē'rak), β *Andromedæ*, "the girdle," or "the loins."

A yellow star culminating Nov. 28th.

MĪZAR (mīzär or mē'zär), ζ *Ursæ Majoris*, "a girdle or apron." Situated in the tail of the Great Bear. Brilliant white in color. It is approaching the earth at the rate of nineteen miles per second. It culminates May 28th.

MULIPHEN, γ *Canis Majoris*.

Situated in the neck of the Greater Dog. It culminates Feb. 26th.

MUPHRID (mū'-frid), η *Boötis*, "the solitary star of thelancer." Situated in the calf of the left leg of the Herdsman. Pale yellow in color. It culminates June 4th.

MURZIM or MIRZAM (mer-zäm'), β *Canis Majoris*, "the announcer" or "the roarer."

Situated in the Greater Dog's left fore paw. A white star culminating Feb. 5th.

MUSCIDA, o *Ursæ Majoris*, "the muzzle."

Situated in the nose of the Great Bear.

NEKKAR, or NAKKAR (nak'-kär), β *Boötis*, "the herdsman," the Arab name for the whole constellation.

Situated in the head of Boötes. A golden yellow star which culminates June 20th.

NAOS (nā'-os), ζ *Argus*, "the ship."

Situated in the stern of the Ship. It culminates Mar. 3d.

NASHIRA, γ *Capricorni*, "the fortunate one, or the bringer of good tidings."

Situated in the tail of the Sea-Goat. It culminates Oct. 3d.

NIHAL, β *Leporis*.

Situated in the right foot of the Hare. Deep yellow in color. It culminates Jan. 23d.

NODUS SECUNDUS, δ *Draconis*, "the second of the four knots or convolutions."

Deep yellow in color. It culminates Aug. 24th.

NUNKI, σ *Sagittarii*, "the star of the proclamation of the sea," or SADIRA (sad'-ē-ra), "the ostrich returning from the water."

Situated in the upper part of the Archer's left arm. It culminates Aug. 17th.

PHAD, PHEC'-DA, or PHAED (fā'-ed), γ *Ursæ Majoris*, "the thigh" (of the bear).

Topaz yellow in color. It is approaching the earth at the rate of sixteen miles per second. It culminates May 4th.

PHAET or PHACT, α *Columbæ*.

Situated in the heart of the Dove. It culminates Jan. 26th.

PHERKAD (fer'-kad), γ *Ursæ Minoris*, "the calf."

Situated in the right fore leg of the Little Bear.

PO-LÁ-RIS, α *Ursæ Minoris*, "the pole star."

Situated in the tip of the Little Bear's tail. Topaz yellow in color. It is receding from the earth at the rate of sixteen miles per second.

POL'-LUX, β *Geminorum*, Ovid's "Pugil," the pugilist of the two brothers.

Situated in the head of Pollux. An orange-colored star which is receding from the earth at the rate of one mile per second. It culminates Feb. 26th. The Century Dictionary gives the color of Pollux as very yellow.

PORRIMA (por'-i-mä), γ *Virginis*, Latin name for "a goddess of prophecy."

Situated in the Virgin's left arm. It culminates May 17th.
PRO'-CY-ON, α *Canis Minoris*, "the foremost dog." A yellowish-white star. It is approaching the earth at the rate of six miles per second. It culminates Feb. 24th. It is situated in the right side of the Lesser Dog. Dr. Elkin gives its distance as 12.3 light years, and its proper motion as 13.9 miles per second.

PROPUS (prō'-pus), η *Geminorum*, "the forward foot."

Situated in the northern foot of Castor.

RASALAS (ras'-a-las), μ *Leonis*, "the lion's head toward the south."

Situated in the Sickle, close to the Lion's right eye. An orange-colored star. It culminates Apr. 1st. Alshemali and Borealis are other names for this star.

RAS ALGETHI (räs-al-ge′-thi), α *Herculis*, "the kneeler's head."
Orange red in color. It culminates July 23d.

RAS′-AL-HĀG′-UE, α *Ophiuchi*, "the head of the serpent charmer."

A sapphire-hued star. It is receding from the earth at the rate of twelve miles per second. It culminates July 28th.

RASTABAN (räs-ta-bän′), β or γ *Draconis* "the dragon's head," or "the head of the basilisk."

A yellow star culminating Aug. 3d. This star also called Alwaid (al-wīd′) "the sucking camel-colts." The three stars near it are included in this appellation.

REG′-U-LUS, α *Leonis*, diminutive of the earlier Rex.

Situated in the handle of the Sickle, and the right fore paw of the Lion. It is flushed white in color, and is approaching the earth at the rate of five miles per second. It culminates April 6th. According to Dr. Elkin it is 35.1 light years distant, and has a proper motion of 8.5 miles per second.

RIGEL (ri′-jel), β *Orionis*, "the [left] leg of the Jabbah, or giant."

A bluish-white star, which is receding from the earth at the rate of ten miles per second. It culminates Jan. 20th. This star is sometimes called Algebar (al′-je-bär).

ROTANEV (rot′-a-nev), β *Delphini*, from Venator, assistant to Piazzi, his name reversed.

It culminates Sept. 15th.

RUCHBA, ω *Cygni*, "the hen's knee."

A pale red star.

RUCHBAH, or RUCBAH, δ *Cassiopeiæ*, "the knee."

Situated in the left knee of Cassiopeia. It culminates Dec. 2d.

RUKBAT, α *Sagittarii*, "the archer's knee."

Situated in the left fore foot of the Archer. It culminates Aug. 24th.

SABIK, η *Ophiuchi*.

A pale yellow star in the left leg of the Serpent Bearer. It culminates Aug. 21st.

SADACHBIA (sād-ak-bē′-yä), γ *Aquarii*, "the luck star of hidden things."

Greenish in color and situated in the water jar of Aquarius. It culminates Oct. 16th.

SAD AL BARI, λ and μ *Pegasi*, "the good luck of the excelling one."

Situated close to the fore legs of Pegasus.

SADAL MELIK (säd-al-mel′-ik), or RUCBAH, α *Aquarii*, "the lucky star of the king."

A red star situated in the right shoulder of Aquarius. It culminates Oct. 9th.

SADALSUND, or SADALSUUD (säd-al-sö-öd), β *Aquarii*, "the luckiest of the lucky."

Pale yellow in color. Situated in the left shoulder of Aquarius. It culminates Sept. 29th.

SADATONI (sad-a-tō′-ni), ζ *Aurigæ*.

One of the three stars known as "the kids." Orange color.

SADR (sadr), or SADIR (sā′-dēr), γ *Cygni*, "the hen's breast."

This star is approaching the earth at the rate of four miles per second. It culminates Sept. 11th.

SAIPH (sā-if′), κ *Orionis*, "the sword of the giant."

Situated in Orion's right knee. It culminates Jan. 27th.

SARGAS, θ *Scorpii*.

A red star situated in the tail of the Scorpion. It culminates July 27th.

SCHEAT (she′-at), or Menkib, β *Pegasi*, "the upper part of the arm."

Situated in the left fore-leg of Pegasus. It is deep yellow in color, and is receding from the earth at the rate of four miles per second. It culminates Oct. 25th.

SCHEMALI, see Deneb al schemali, ι *Ceti*.

SEGINUS (se-jī′nus), γ *Boötis*, from Ceginus of the constellation, possibly.

Situated in the left shoulder of Boötes. It culminates June 13th.

SHAULA (shâ′-lä), λ *Scorpii*, "the sting."

In the tip of the Scorpion's tail.

SHEDAR, SCHEDIR, or SHEDIR, α *Cassiopeiæ*, "the breast," or from El Seder, "the sedar tree," a name given to this constellation by Ulugh Beigh.

Pale rose in color. It culminates Nov. 18th.

SHELIAK, or SHELYAK (shel´-yak), "a tortoise," β *Lyræ*, Arabian name for the constellation.

A very white star culminating Aug. 17th.

SHERATAN (sher-a-tan´), β *Arietis*, "a sign," or "the two signs." Situated in the Ram's horn. A pearly white star culminating Dec. 7th.

SIR´-I-US, α *Canis Majoris*, "the sparkling star or scorcher." Situated in the mouth of the Great Dog. Brilliant white in color. The brightest of the fixed stars. It culminates Feb. 11th.

SITULA (sit´-ū-lā), κ *Aquarii*, "the water jar or bucket." Situated in the rim of the Water Jar.

SKAT, or SCHEAT, δ *Aquarii*, "a wish," or possibly it means a "shin bone."

Situated in the right leg of Aquarius.

SPÏ´CA, α *Virginis*, "the ear of wheat or corn" (held in the Virgin's left hand).

A brilliant flushed white star, which is approaching the earth at the rate of nine miles a second. It culminates May 28th.

SUALOCIN, or SVALOCIN (sval´-ō-sin), Nicolaus reversed, α *Delphini*.

A pale yellow star culminating Sept. 15th.

SULAFAT, or SULAPHAT (sö´-lä-fät), "the tortoise," γ *Lyræ*.

Arabian title for the whole constellation. It is bright yellow in color, and culminates Aug. 19th.

SYRMA, ι *Virginis*; this name used by Ptolemy to designate this star in the train of the Virgin's robe.

TALITA (tä´-lē-tä), κ or ι *Ursæ Majoris*, "the third vertebra." Situated in the right fore paw of the Great Bear. Topaz yellow in color.

TANIA BOREALIS, λ *Ursæ Majoris*.

TANIA AUSTRALIS, μ *Ursæ Majoris*, a red star.

These stars are situated in the right hind foot of the Great Bear. The former star culminates Apr. 8th.

TARAZED (tar'-a-zed), γ *Aquilæ*, "the soaring falcon," part of the Persian title for the constellation.

Situated in the body of the Eagle. A pale orange star, culminating Aug. 31st.

TEGMENI, ζ *Cancri*, "in the covering."

A yellow-colored star.

TE'-JAT, μ *Geminorum*.

THU'-BAN or (thō-ban'), α *Draconis*, "the dragon," the Arab title for the constellation.

Situated in one of the Dragon's coils. It is pale yellow in color, and culminates June 7th.

UNUK AL HAY or UNUKALHAI (ū'-nuk-al-hä'-i), α *Serpentis*, "the neck of the snake."

A pale yellow star which is receding from the earth at the rate of fourteen miles a second. It culminates July 28th.

VĔ'GA, or WEGA, α *Lyræ*, "falling," *i.e.*, the falling bird, "the harp star."

A beautiful pale star sapphire in color. It is approaching the earth at the rate of nine miles a second. It culminates Aug. 12th.

VINDEMIATRIX, ε *Virginis*, "the vintager or grape gatherer."

Situated in the Virgin's right arm. A bright yellow star culminating May 22d.

WASAT (wä'-sat), δ *Geminorum*, "the middle."

Situated in the body of Pollux. Pale white in color. It culminates Feb. 19th.

WESEN, δ *Canis Majoris*, "the weight."

A light yellow star in the right side of the Great Dog. It culminates Feb. 17th.

YED PRIOR (yed), δ *Ophiuchi*, "the hand," "the star behind or following."

Deep yellow in color. It culminates July 7th. It is in the left hand of the Serpent Bearer.

YED POSTERIOR, ε *Ophiuchi*, "the hand."

A red star culminating July 8th.

YILDUM, δ *Ursæ Minoris.*
Situated in the tail of the Little Bear. A greenish-hued star culminating Aug. 12th.

ZANIAH, η *Virginis.*
Situated in the Virgin's left shoulder.

ZAURAK (zâ'-rak), γ *Eridani*, "the bright star of the boat."
A yellow star.

ZAVIJAVA (zav-ija'-va), β *Virginis*, "angle or corner," "the retreat or kennel of the barking dog."
Situated on the Virgin's left wing. A pale yellow star culminating May 3d.

ZOSMA (zōs'-ma), δ *Leonis*, "a girdle."
Situated at the root of the Lion's tail. A pale yellow star which is approaching the earth at the rate of nine miles a second. It culminates Apr. 24th. This star is also called Duhr, and sometimes Zubra.

ZUBENAKRAVI (zöben-ak'-ra-vi or -bi), γ *Scorpii*, "the claw of the Scorpion." A red star.

ZUBEN ELGENUBI (zö-ben-el-jen-ū'-bi), α *Libræ*, "the southern claw" (of the Scorpion).

A pale yellow star culminating June 17th. This star is also called Kiffa Australis.

ZUBEN ESCHAMALI (zö-ben-es-she-ma'-li), β *Libræ*, "the northern claw."

A pale emerald color, a very unusual color for a star. It is approaching the earth at the rate of six miles a second and culminates June 23d. This star is also known as "Kiffa Borealis."

In the compilation of the foregoing list, the author has been greatly assisted by Allen's "Star Names and their Meanings."

TABLE SHOWING THE STARS OF THE FIRST AND SECOND MAGNITUDE RISING IN THE EASTERN SKY AT NINE O'CLOCK P.M. ON THE DATES SPECIFIED.

DATE		NAME OF STAR	CONSTELLATION
January	1	Regulus, 1st.	Leo.
"	8	Alphard, 2d.	Hydra.
"	11	Cor Caroli.	Canes Venatici.
February	20	Arcturus,	1st. Boötes.
March	1	Spica, 1st.	Virgo.
"	5	Gemma, 2d.	Corona Borealis.
April	1	Vega, 1st.	Lyra.
"	20	Ras Alhague, 2d.	Ophiuchus.
"	22	Deneb, 2d.	Cygnus.
May	9	Antares, 1st.	Scorpius.
"	26	Altair, 1st.	Aquila.
June	5		Delphinus.
July	17	Algenib, 2d.	Perseus.
August	6	Algol.	Perseus.
"	21	Capella, 1st.	Auriga.
"	"	Hamal, 2d.	Aries.
"	27	Fomalhaut, 1st.	Piscis Australis.
September	13		The Pleiades in Taurus.
October	2	Aldebaran, 1st.	Taurus.
"	26	Bellatrix, 2d.	Orion.
"	30	Castor, 2d.	Gemini.
"	"	Betelgeuze, 1st.	Orion.
November	4	Pollux, 1st.	Gemini.
"	"	Rigel, 1st.	Orion.
"	27	Procyon, 1st.	Canis Minor.
December	4	Sirius, 1st.	Canis Major.
"	8	Phaet, 2d.	Columba.
"	14		The Bee Hive in Cancer.
"	16		The head of Hydra.